Johnny Ball

Mathmagicians
WAVES
FREQUENCY
LIGHT
GRAVITY
GOOGOL
KILOGRAM
DISTANCE
WEIGHT
CELSIUS
NUMBERS
CURRENT
DENSITY
AREA
CHARTS
DK

Penguin Random House

Author Johnny Ball
Senior editor Ben Morgan
Senior art editor Claire Patané
Editors Wendy Horobin, Elinor Greenwood, Chris Woodford, Carrie Love, Fleur Star, Joe Harris
Designers Laura Roberts-Jensen, Sadie Thomas, Hedi Hunter, Clémence Monot, Lauren Rosier
Picture researcher Rob Nunn
Indexer Chris Bernstein
Production editor Clare McLean
Production controller Claire Pearson
Jacket designers Karen Shooter, Akiko Kato
Jacket editor Mariza O'Keeffe
Publishing manager Bridget Giles
Art director Rachael Foster
Creative director Jane Bull
Publisher Mary Ling

Consultant Dr Jon Woodcock

REVISED EDITION
Senior editors Fleur Star, Sreshtha Bhattacharya
Senior art editors Spencer Holbrook, Ranjita Bhattacharji
Assistant editor Sheryl Sadana
Assistant art editor Kshitiz Dobhal
Senior DTP designer Harish Aggarwal
DTP designer Sachin Gupta
Jacket designer Surabhi Wadhwa
Jacket assistant Claire Gell
Jacket design development manager Sophia MTT
Producer, pre-production Gillian Reid
Producer Vivienne Yong
Managing editors Linda Esposito, Kingshuk Ghoshal
Managing art editors Philip Letsu, Govind Mittal
Publisher Andrew Macintyre
Publishing director Jonathan Metcalf
Associate publishing director Liz Wheeler
Design director Stuart Jackman

First published in hardback in Great Britain in 2009
This paperback edition first published in 2016 by
Dorling Kindersley Limited
80 Strand, London WC2R 0RL

2 4 6 8 10 9 7 5 3 1
001 – 290013 – January/2016

A CIP catalogue record for this book is available from the British Library.

Paperback edition ISBN: 978-0-24124-357-2

Printed and bound in China

A WORLD OF IDEAS:
SEE ALL THERE IS TO KNOW
www.dk.com

“My previous Dorling Kindersley book was called *Think of a Number* (or *Go Figure!* in the US). It told the amazing story of where numbers came from and showed how they can be surprising, mischievous, and fun. It also dipped into the weird and wonderful world of modern mathematics.

But numbers are no good unless we use them, and that's what this book is all about. We use numbers not just to count but to *measure*. Without measuring, we wouldn't be able to plan, design, or build. We wouldn't be able to explore the world or make scientific breakthroughs. Now science can be complicated, but I hope to show you how maths can make it magically simple to understand.

In this book I will take you back in time to the very beginnings of maths and measuring. I will introduce the *mathmagicians* – the people throughout history who have used the wizardry of numbers to make sense of the world and unravel the secrets of the Universe. It's a story that takes us right up to the present day and reveals the imaginative ways we measure absolutely everything today.

I hope you enjoy the book and I hope it will help you to love maths and science as much as I do. If it works, then perhaps, like me, you'll go through life always wanting to know and understand more. That's how I think we should all live our lives.”

0.5

360

14.7

1000

90°

18

52

40°

12

CONTENTS

N
W
E
S

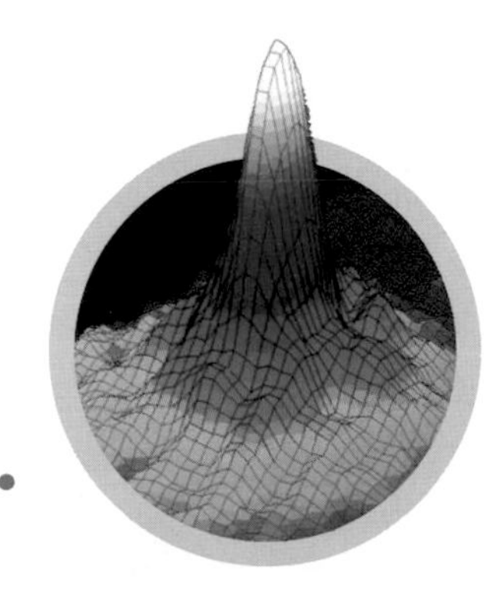

Imagine what the world would be like without **MEASUREMENTS...**

THE DAILY

Business & Finance

PRICE OF PETROL RISES

The price of petrol rose this week to £1 for a short squeeze of a petrol pump, £1.50 for a medium squeeze, and £2 for a long squeeze. Disputes continue to flare up at filling stations as drivers and petrol pump attendants argue over the meaning of short, medium, and long. Meanwhile, in Preston a farmer drove a milk tanker into a filling station and completely filled it with a single long squeeze costing only £2, leaving the filling station empty.

The long and winding road

ROAD RAGE

by **Ben D. Lane**

A row has erupted over a brand new road that has kinks all the way along it. Chief engineer Mac Adam explained the problem: "We don't know exactly how long roads have to be, so we guess. If we guess right, you get a straight road. If we guess wrong, we have to put bends in to make the road fit between the towns. When we get it really wrong, we stick some hills in as well."

Weather

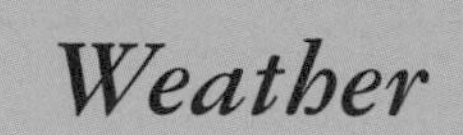

Tomorrow: lots of rain but it's hard to say how much.

The day after tomorrow: sunny and sort of warmish.

The day after the day after tomorrow: quite hot actually.

The day after the day after the day after tomorrow: scorchio!

WORLD'S TALLEST BUILDING?

by **Bill Ding**

In a bid to find out which is the world's tallest building, plans are being made to move the ten tallest-looking skyscrapers to one place so they can stand side by side. Each skyscraper will be carefully dismantled, shipped to the USA, and then rebuilt. Once the winner is known, the buildings will be dismantled, shipped home, and rebuilt all over again. Governments are still arguing over who is going to foot the bill – which is also expected to be sky-high.

PLANET

LATEST EDITION (UNLESS YOU CAN FIND A LATER ONE)

SPORT & LEISURE

They hope it's all over!

by Johnny Ball

A soccer match between England and Brazil – thought to be the longest match ever played – shows no sign of coming to an end. With no way of measuring time, no one has any idea how long the match has been running, when it should end, or even when to call half time.

The players were in their twenties at the start of the match but most are now too old to get around without wheelchairs or walking sticks. One especially old player has threatened to burst the ball with his stick so he can go home.

During the course of the match, around 3000 spectators have died of old age and another 1500 have died of boredom. The current score is Brazil 75,789, England 76,100.

England's three youngest players can still get around the pitch without wheelchairs.

Fisherman Catches **HUGE** fish

Ivor Hook with his big fish.

Fisherman Ivor Hook yesterday caught a really big fish. Hook has caught huge fish before, but he says this one is really, *really* huge. He isn't sure that it's his biggest, though, because he ate all his previous catches and so can't compare. He thinks the new fish might be heavier than he is, but he isn't sure about that either as he doesn't know his own weight.

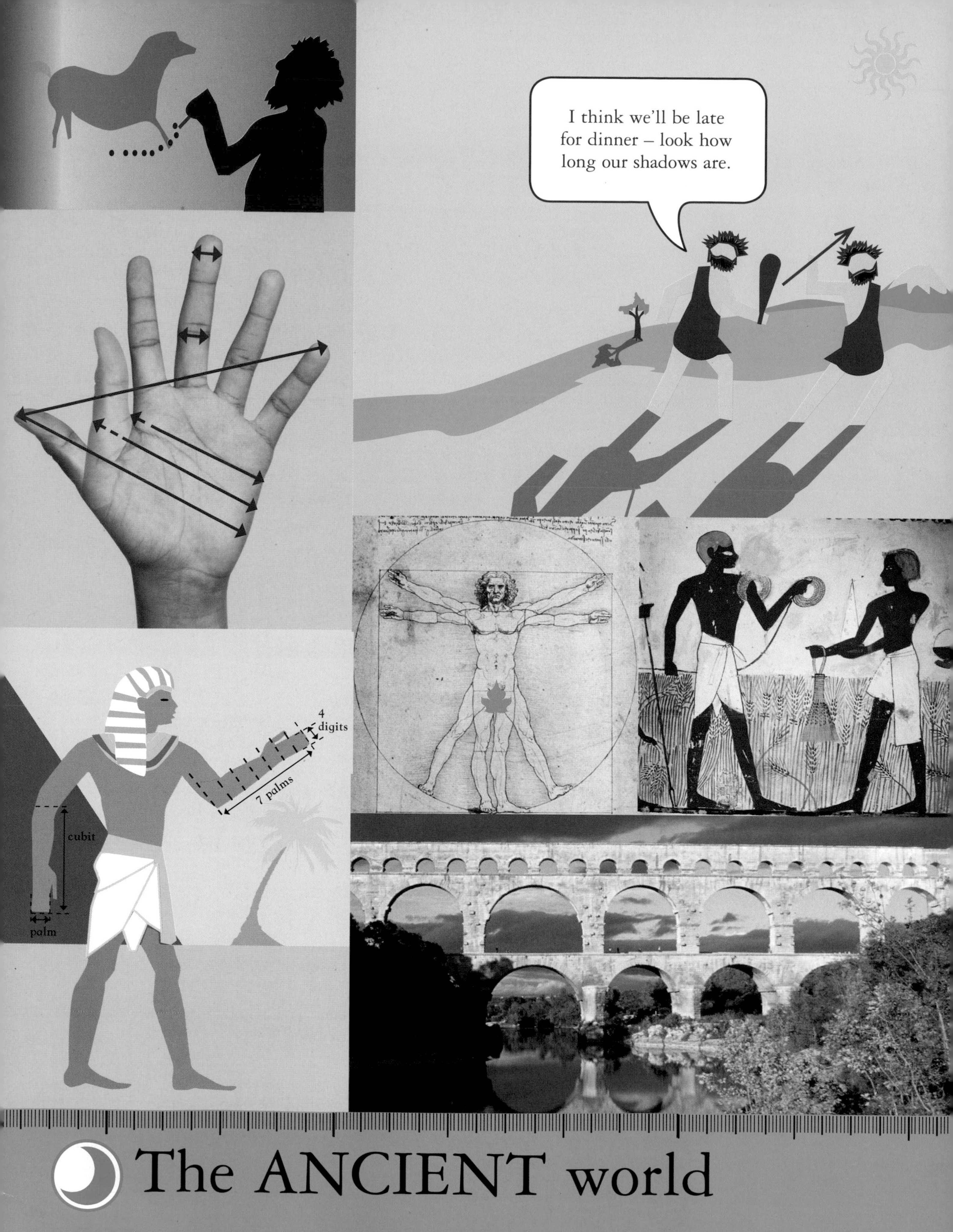

The ANCIENT world

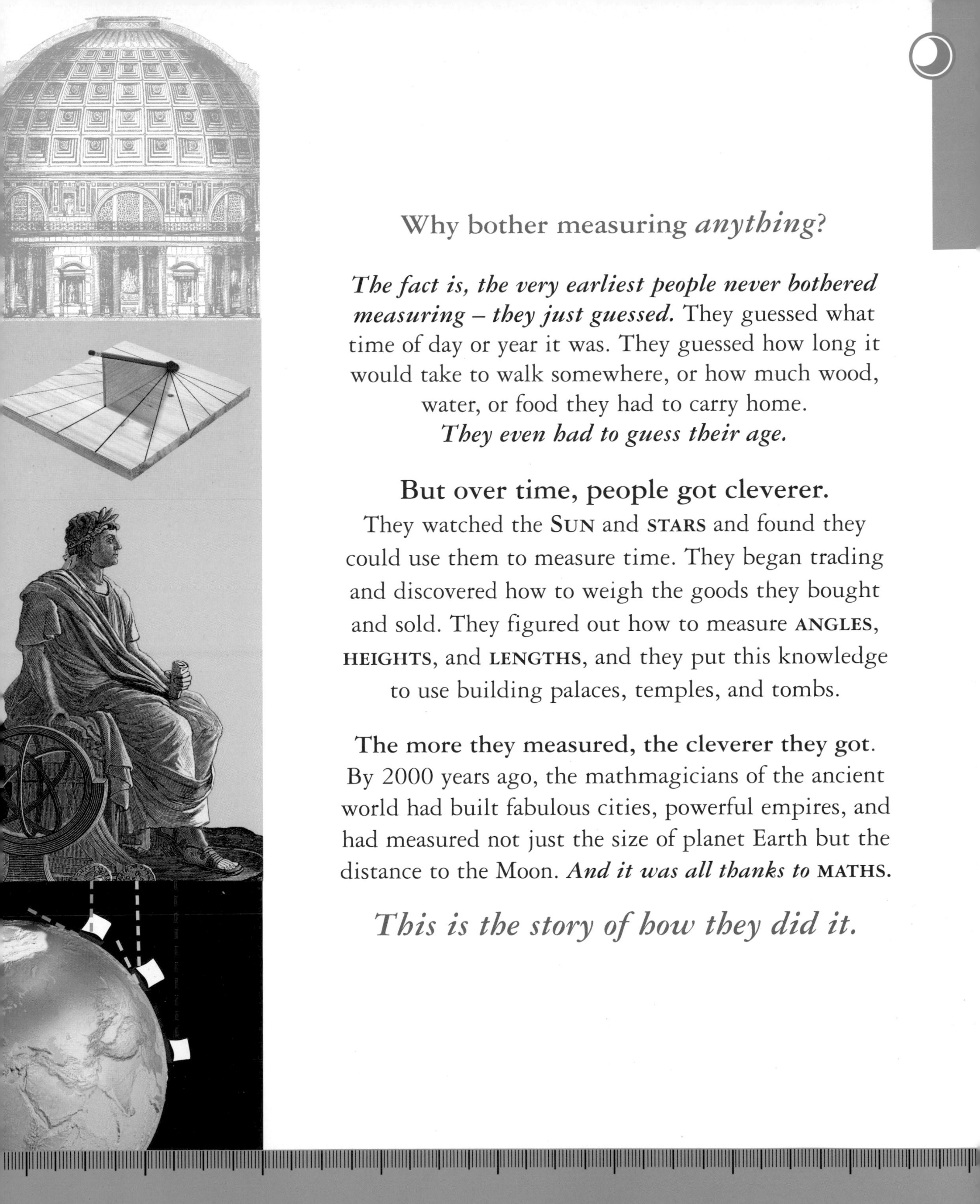

Why bother measuring *anything*?

The fact is, the very earliest people never bothered measuring – they just guessed. They guessed what time of day or year it was. They guessed how long it would take to walk somewhere, or how much wood, water, or food they had to carry home.
They even had to guess their age.

But over time, people got cleverer.
They watched the **SUN** and **STARS** and found they could use them to measure time. They began trading and discovered how to weigh the goods they bought and sold. They figured out how to measure **ANGLES**, **HEIGHTS**, and **LENGTHS**, and they put this knowledge to use building palaces, temples, and tombs.

The more they measured, the cleverer they got.
By 2000 years ago, the mathmagicians of the ancient world had built fabulous cities, powerful empires, and had measured not just the size of planet Earth but the distance to the Moon. ***And it was all thanks to*** **MATHS.**

This is the story of how they did it.

MOONS and *months*

Ancient people measured long distances not in metres or miles but in the time it took to walk. A distant river or mountain might be "two days' walk" away, for instance, or it might be near enough to reach "by sundown". By watching the Sun's movement across the sky and noting the length of the shadows it cast, people had a rough idea of how much daylight they had left.

I think we'll be late for dinner – look how long our shadows are.

New moon

Waxing crescent

To measure longer periods of time, people had to count the days. Counting seems easy to us, but the very earliest people weren't good at it.

Yuk! These strawberries are nowhere near ripe, let's come back the moon after next.

Counting full moons would have come in handy. Imagine the ancient people came across a fruit tree or a bush covered in berries during their wanderings. If the fruit wasn't ripe they might have decided to wait until the moon reached a certain point in its cycle and then return. They may even have known that each of the four seasons lasts about three full moons, which gave them a rough way of measuring the year.

There are 12 full moons in most years

How long is a month?

A month is roughly the length of time it takes the Moon to orbit Earth once. We see the Moon changing shape over the course of the month because it's position relative to the Sun and Earth keeps changing, causing us to see a varying amount of its sunlit side and dark side. It takes 27.3 days for the Moon to orbit Earth exactly once. This is known as the sidereal month. But the time between two new or two full moons is slightly more: 29.5 days. The reason for the difference is that Earth is moving around the Sun while the Moon is moving around Earth. With each month, the Moon has to travel nearly two days further than one whole orbit before it becomes a new moon again.

Position after 27.3 days

Position after 29.5 days

What direction does the shadow progress across the Moon's face during the lunar cycle?

Long before people had clocks or watches, our ancient Stone-Age ancestors could measure time by counting days or by watching the Sun, Moon, and stars. Time was one of the very first things that people started measuring.

They counted on their fingers, and since they only had ten of those (including the thumbs) they found it very difficult to count larger numbers. But they had another way of keeping track of the long periods we now call weeks and months: they watched the Moon. Our early ancestors saw how the Moon gradually changes from a thin crescent to a large white disc – a full Moon – as the days progress.

Later, people discovered they could count past ten if they stopped using their fingers and instead used some other memory aid. Some people carved notches in a tree. Others daubed spots of paint onto cave walls or tied knots in string. They soon found that it takes about 30 days for the Moon to go through its cycle – a length of time we now call a month. The ancients also discovered there are 12 months in a year. By multiplying they measured the length of a year in days: 30 × 12 = 360. The answer is wrong, of course, but it was near enough for Stone Age people. As we'll find out on the next page, it wasn't until people started using the Sun and stars to measure the year that they got it right.

and 29.5 days between each one.

When the Sun eats the Moon

As the Moon travels around Earth over and over again, it sometimes happens to pass right through Earth's shadow. When this happens we see a lunar eclipse – the Moon turns a dark reddish colour as it plunges into the dark zone behind Earth. You might wonder why we don't see an eclipse every month. The reason is that the Moon's orbit is tilted. It usually flies just above or just under Earth's circular shadow, but once every few months or so it flies at exactly the right height to hit the shadow and we get a magical total eclipse.

The Moon orbits Earth on a plane that's tilted relative to the plane of Earth's orbit around the Sun.

Right to left from the Northern Hemisphere and left to right from the Southern Hemisphere.

Measuring the year by counting moons was good enough for early Stone Age people because they never needed to know the exact date. But about 10,000 years ago, people had to smarten up their act. Something amazing happened that made it vitally important to measure the year precisely.

The earliest people never needed to know exactly what time of year it was because they led simple lives, wandering from place to place and gathering all the food they needed from the wild. They had no calendars, never knew the date, and couldn't celebrate their birthdays. But around 10,000 years ago, people in the Middle East discovered they could grow wheat instead of gathering it. These first farmers could at last stay in one place, and their settlements grew to become the world's first towns. To get the best possible harvest, they had to sow at just the right time, so they became experts at measuring the year.

The farmers of ancient Egypt had to grow their crops in winter because the River Nile flooded every summer, covering the fields with water. The Egyptians noticed that the star Sirius made its first appearance in the night sky in early summer each year, before the floods. So they measured the year by counting the days after Sirius rose and discovered that a year is 365 days long.

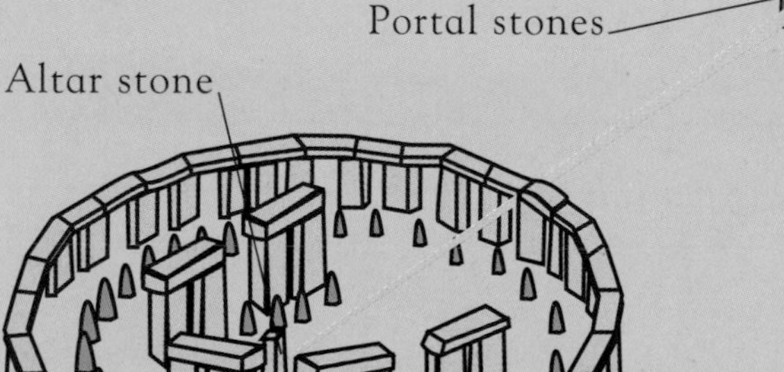

While the Egyptians were building their pyramids and temples, mathematician-priests in Europe were also building temples to the Sun that helped them work out the date. Stonehenge in England was designed to track the Sun's movement and reveal when midsummer's day arrives. Only on that day did a beam of sunlight from the rising midsummer Sun pass through two "portal stones" outside the main circle to strike a central "altar stone".

By tracking the Sun and stars, ancient people

How long is a year?

We now know that a year is actually slightly more than 365 days long. Earth takes 365.2425 days to orbit the Sun once, which isn't a whole number of days. So to compensate, we add an extra day (the 29th of February) once every 4 years to make the year 366 days long – a "leap year". And to keep the calendar really accurate, every hundredth year is not a leap year. Is it good or bad luck to be born on February 29th? You only get one true birthday every four years, but think of this: after you've lived 60 years, you're still only 15!

How many times a year does the Sun set at the North Pole?

Seasons in the SUN

The clever Egyptians also knew how to measure the year by tracking the position of the rising Sun. Measuring the year this way was such an important job that the Sun came to be worshipped as a god, and the mathmagicians who could track its movement and work out the date became priests. At Karnak in the south of Egypt, the priests had a fabulous temple built in honour of the Sun. A row of enormous columns were positioned so that on midwinter's day each year, the rising Sun would send a shaft of light down the aisle between the columns and straight into the heart of the temple.

How many days until Christmas?

The native Mayan people of Central America also discovered how to grow crops and so became experts at measuring the year. Like the Egyptians and Europeans, they figured out that a year is 365 days long and they built temples in honour of their sacred calendar and sun-god. The pyramid at Chichen Itza in Mexico has four staircases of 91 steps and a single platform on top, making 365 in total – the length of a year. The Mayans were brilliant at maths but were also deeply superstitious. To appease the gods and protect their crops, they performed human sacrifices, tearing the beating hearts out of victims' bodies while still alive.

worked out that a year is *365 days* long.

What causes seasons?

Planet Earth isn't quite upright – it spins round at a tilt. The four seasons of spring, summer, autumn, and winter happen because this tilt makes different parts of the planet lean towards the Sun or away from it during the year-long orbit. In the Northern Hemisphere, summer occurs when the North Pole leans towards the Sun, making northern countries sunnier and giving them longer days. When the North Pole leans away from the Sun, it's winter in the north and summer in the south.

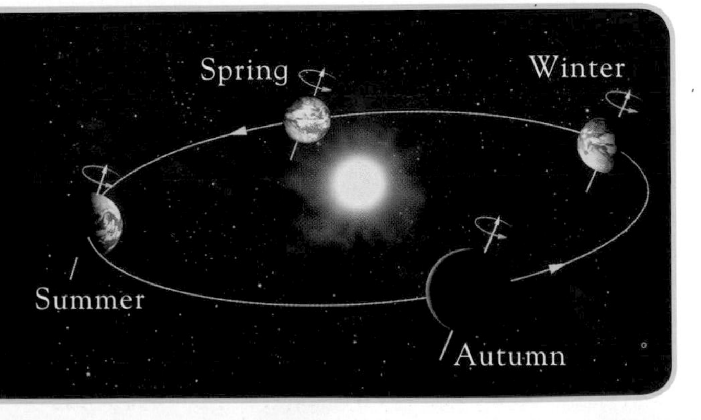

Once. The Sun rises and sets only once a year at the poles, resulting in 6 months of daylight followed by a 6-month night.

The RIGHT angles

As farming spread and civilization flourished, people's mathematical skills advanced. The Egyptians used their measuring skills to plan and construct vast tombs with perfectly square bases and triangular sides – the pyramids. To build them, they had to become experts at measuring and marking out angles.

The most important angle used by the Egyptians was the right angle, which is 90° (one quarter of a whole circle). Right angles create square corners, which are vital in construction.

The Great Pyramid

Built in 2560 BCE, the Great Pyramid of Khufu was the world's tallest building for nearly 4000 years. The angle of the slope is a consistent 52° all the way up. The Egyptians got it right by setting the stones 22 fingers' width nearer the centre with each 28 fingers' width increase in height.

The ***Great* Pyramid** contains enough stone to build a wall *2 metres (6.5 feet) tall and 18 cm (7 in) wide* all the way from CAIRO to the NORTH POLE.

Hurry up with my tomb!

The Egyptians used at least three tools to make sure all the angles were right.

Every block of stone was cut by hand. The corners had to be right angles so the blocks would stack neatly. Builders checked each corner with a tool called a mason's square.

To make sure the top of each block was exactly level, the builders placed a triangular tool on top and checked whether the attached weight hung in the middle.

The sides of the blocks had to be at right angles with the ground. The builders checked this with a plumb line – a weight hanging on a string.

THE GREAT PYRAMID OF KHUFU is the only surviving member of the *seven wonders* of the world

PLANNING THE BASE

One of the trickiest problems for the Egyptians would have been making sure the base of the pyramid was perfectly square, with right-angled corners. The corners could have been marked with pegs and ropes using the technique shown below. The ground also had to be made perfectly flat. This could have been done by making water-filled trenches and then levelling the ground to match the water. Afterwards the trenches were filled back in.

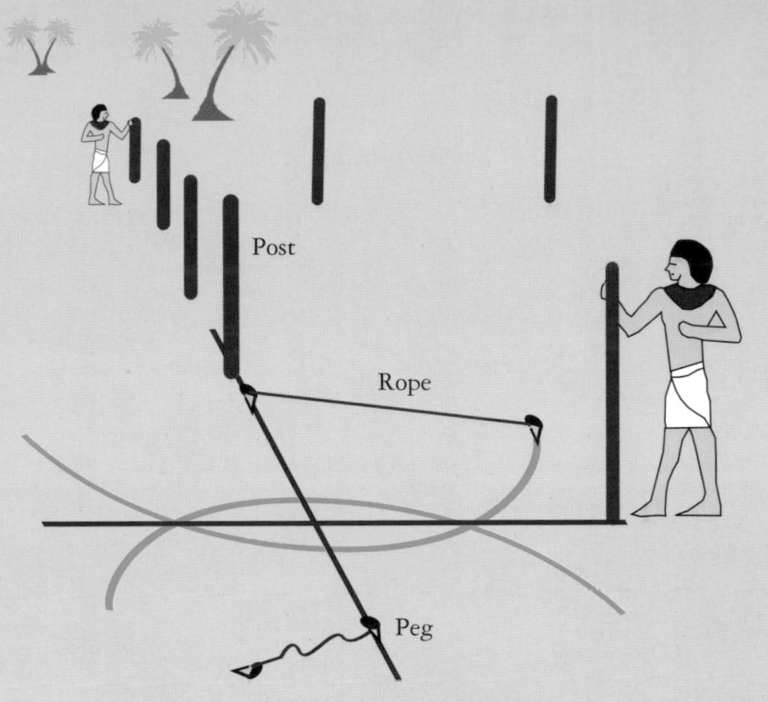

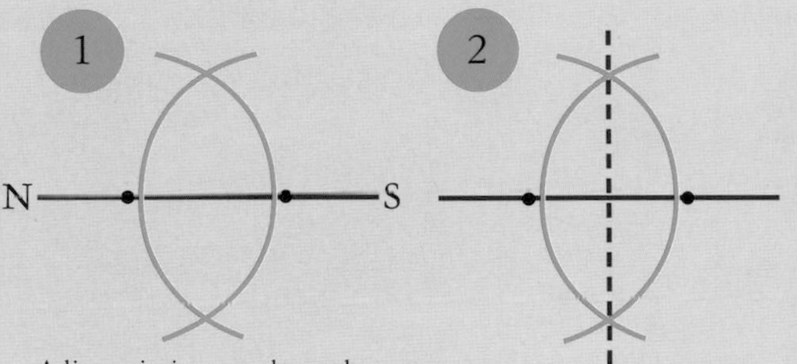

A line pointing exactly north–south was drawn and two points were marked on it. With these points as centres, two circular arcs were drawn to overlap each other.

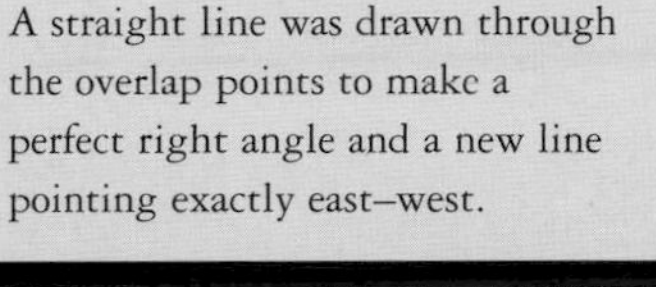

A straight line was drawn through the overlap points to make a perfect right angle and a new line pointing exactly east–west.

As well as making sure the corners were square, the Egyptians had to get the sides perfectly straight. To do this they might have placed posts in the ground and then looked along them to make sure they lined up.

FINDING NORTH

Viewed from space by satellite, the pyramids line up exactly with the points of the compass. Yet they were built thousands of years before magnetic compasses were invented, so how did the builders achieve this amazing feat? The Egyptians were able to locate true north by looking at noon shadows (which always point north) or by viewing the Pole Star. By drawing a line at right angles to the north–south line, they could then locate east and west.

N
W
E
S

Satellite image of the pyramids

PYRAMID FACTS

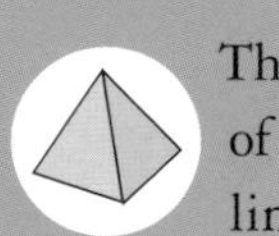

The Great Pyramid is made of 2.3 million blocks of limestone, some weighing as much as 15 tonnes. The blocks fit together so neatly that you can't push a credit card between them.

The pyramids were built before the invention of the wheel. The heavy stone blocks were carried on rafts along the River Nile and then loaded onto sledges to be dragged up specially made stone ramps.

The pyramids were dazzling white when newly built, with a smooth, flat surface that made them impossible to climb. The peak was capped with gold.

Measuring land

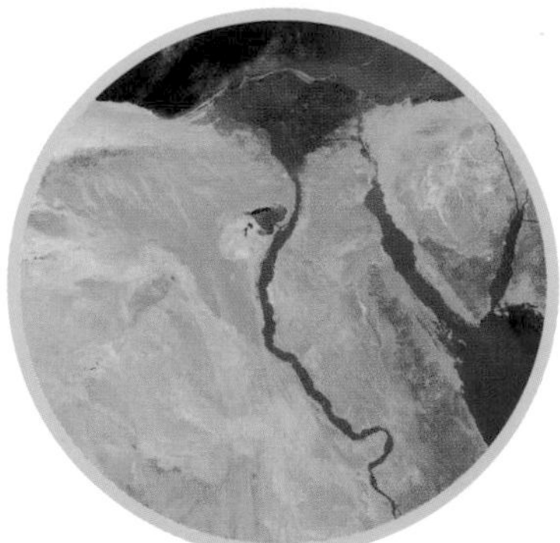
The River Nile brings water to the desert, feeding lush vegetation along its banks.

Egypt lies in the Sahara desert – one of the hottest and driest places on Earth. But thanks to the River Nile, which crosses the desert, there is a thin strip of fertile land that's perfect for farming. For countless centuries, farmers have sown and harvested wheat along the banks of the Nile. The secret of the land's fertility was the river's annual flood. In ancient times, before the Nile was dammed, the river burst its banks every summer and flooded the farmland. When the water drained away, it left a layer of fine mud that was rich in nutrients, fertilizing the soil.

12

13

5

A triangle with sides 5, 12, and 13 units long always forms a right angle.

The Nile's annual floods also washed away the boundaries of fields, so every year the Egyptian farmers had to mark out the land again. This was an important job. The farmers had to know the exact size of their plots because the Egyptian rulers taxed them by the area of their land. To mark out the land into new fields, the farmers used long ropes with regularly spaced knots. They stretched these into triangles, counting the knots along each side to get the shape right, and then pegged out the corners.

5

4

3

The farmers knew that triangles with sides 3, 4, and 5 units long always formed a right angle. So did triangles with sides 5, 12, and 13 units long. By joining two right-angled triangles together, they could make a rectangular plot of known area. It would have been quick and easy work for the farmers to divide up a long stretch of land into rectangles this way, simply by moving one peg at a time and flipping the rope over to make another triangle.

Have we made a right angle yet?

Measuring area

This hieroglyphic shows Egyptian farmers using a knotted rope to measure a wheat field.

Most of the farmers' plots were probably simple rectangles, but what if some areas of farmland were an awkward shape? How could the tax men work out the area of land to tax? With a bit of ingenuity, the Egyptians could have done this with triangles too.

1. Any shape that has straight sides can be divided into a set of right-angled triangles by drawing straight lines across it.

2. It's easy to work out the area of each triangle because a right-angled triangle is simply half a rectangle.

5

7

3. So you multiply the length by the width and divide by two. Then you simply add up all your triangles.

$$5 \times 7 = 35 \qquad 35 \div 2 = 17.5$$

Puzzle

Using the technique above, see if you can work out the area of this four-sided shape, assuming each grey square is one square centimetre. The answer is at the back of the book.

I'm a frayed knot.

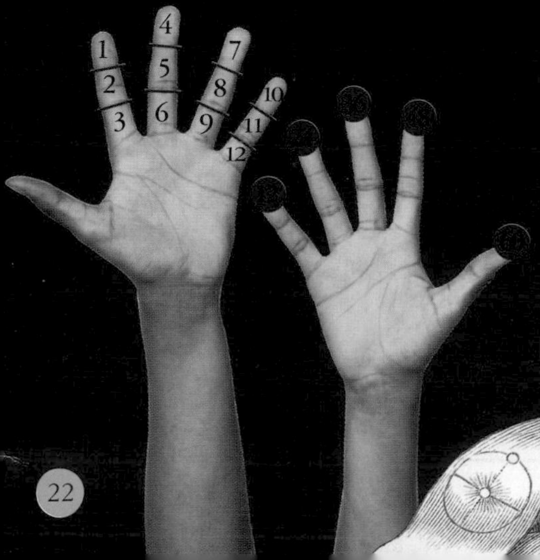

It's all *Greek*

Watching stars, building pyramids, and measuring land helped the Egyptians learn a lot about angles and triangles. Their expertise was handed on to a later civilization – that of ancient Greece. The Greeks found out even more and turned their knowledge of triangles and shapes into a whole new branch of maths: geometry (meaning "Earth measuring").

Measuring angles

The stargazers of ancient Babylonia (now Iraq) noticed that the stars rise in a slightly different position each night and move through a circle over the course of a year. They called the small daily changes "degrees", and since there were about 360 days in a year (according to the Babylonian calendar), they divided the circle into 360 degrees too. Today we still use degrees for measuring angles, which are really just portions of a whole circle.

Minutes and seconds

The Babylonians measured the angles of moving stars with amazing accuracy. They divided each degree into 60 "minutes" and each minute into 60 "seconds". We still use the system today, not just for degrees but for time also. But why divide into 60s rather than 10s or 100s? Perhaps the answer was that the Babylonians counted on finger segments rather than just fingers. By using one hand to tally counts made on the other, they could have used both to count to a total of 60.

Triangles and squares

One of the greatest mathmagicians was a man called Pythagoras. He was fascinated by the right-angled triangles the Egyptians used for measuring land. The Egyptians had discovered you can make right angles by creating triangles with sides 3, 4, and 5 units long or 5, 12, and 13 units long. Pythagoras discovered something else. He drew squares on the sides and found that the areas of the two smaller squares always seemed to add up to the large square. Then he went a step further and proved, using mathematical logic, that the squares on any kind of right-angled triangle must always add up this way. He'd discovered a mathematical law.

Pythagoras made maths into a kind of religion, with himself appointed head priest. His band of devoted followers used secret mathematical codes to identify themselves. They all believed that mathematical patterns lay behind everything from the movement of stars to the sound of music.

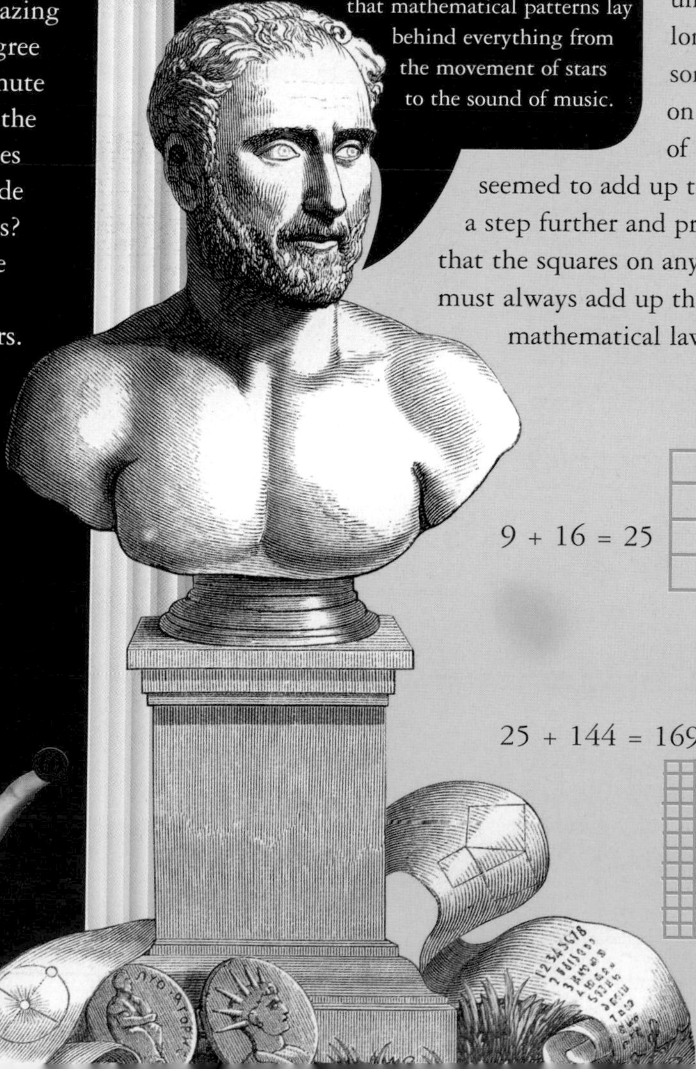

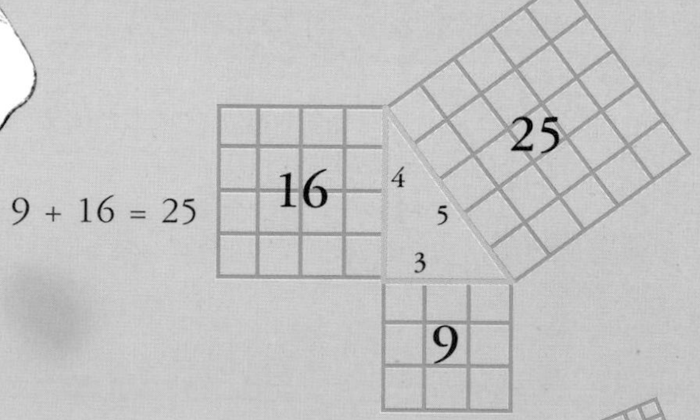

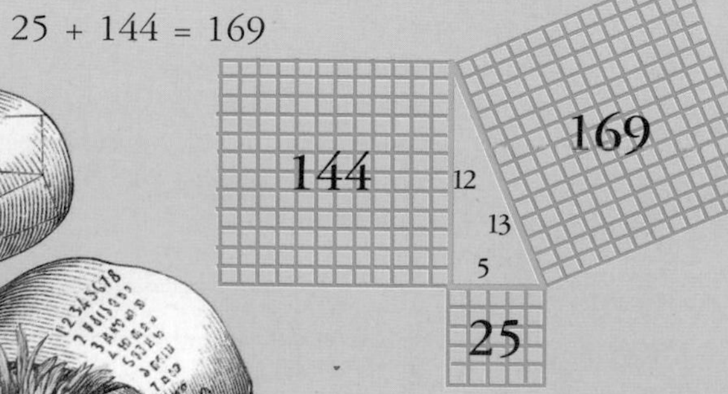

$9 + 16 = 25$

$25 + 144 = 169$

Triangle tricks

Using his knowledge of triangles, a Greek mathematician called Thales thought of a cunning way of measuring how tall things are without having to climb them. Wait until the length of your shadow is the same as your height. At that moment, everything else – from trees to temples – will also have shadows as long as they are tall. Then you can simply measure their shadow to find their heights.

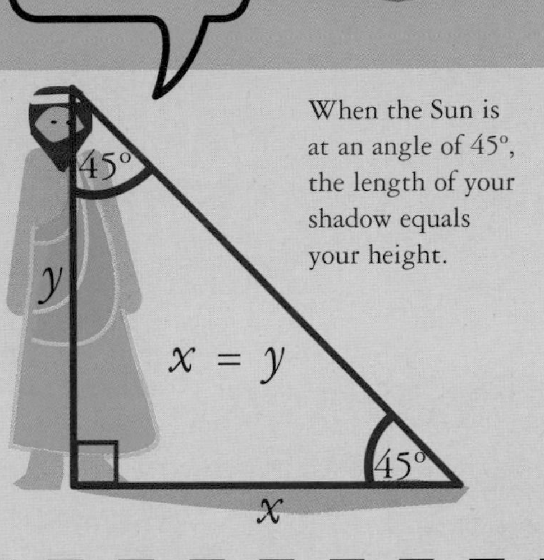

When the Sun is at an angle of 45°, the length of your shadow equals your height.

Thales's trick worked because the Sun, his body, and his shadow formed a special kind of triangle. One corner is a full 90° (a right angle) and the other two are both 45° (half a right angle). The Greeks knew that if two angles of a triangle are equal, then two sides must be equal too. The same triangle is handy for measuring other things. Imagine you want to know how far a ship is from the shore. All you have to do is find a point where the ship is at a right angle to the shore and a point where it's at 45°. The distance between these two points tells you how far the ship is.

When a lumberjack chops down a tree, how near can you stand without being in danger of getting squashed? You need to be at least as far as the tree is tall (but then a bit extra, for safety). If the angle from the ground where you're standing to the top of the tree is 45° or more, you're too close. But if the angle is smaller, you're further than the height of the tree. One rule of thumb is to use a 3, 4, 5 triangle (with the 4 side on the ground), which allows the tree a bit of room to slide.

45°

TIMBER!

I'm off – I forgot my protractor!

45°

Using two triangles

Measuring the height of something when the Sun is shining at 45° is easy, but what if the Sun's rays hit the ground at some other angle? Another Greek mathematician, called Hipparchus, had the answer. He could figure out the height of a pillar by using two shadows: one cast by the pillar and another cast by a smaller object such as a man, whose height is easier to measure. The shadows form two right-angled triangles that are different in size but the same shape.

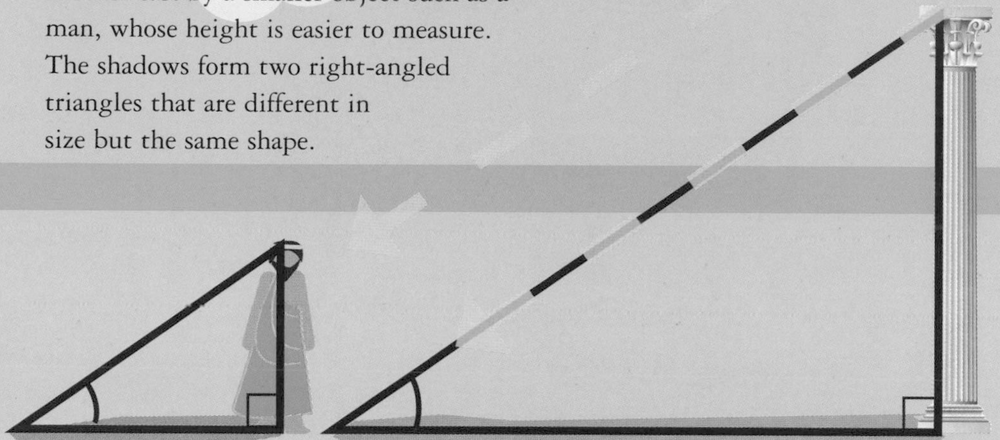

In these triangles the angle on the left is always the same. Hipparchus realized this meant the large triangle is simply a scaled-up version of the small one. So if the man's shadow was twice as long as his height, then the pillar's shadow must be twice its height too. Whatever the angle of the Sun, the man's height divided by his shadow, times the long shadow, will always give the height of the pillar. That's clever.

I, Hipparchus, was the greatest astronomer of ancient Greece. Everyone else stood in my shadow!

TRIGONOMETRY

Hipparchus went even further. He realized that he didn't need two triangles – he could do it all with one. Imagine the length of a man's shadow shrinking as the sun climbs. As the angle of the Sun's rays grows steeper, the man's shadow (*b*) gets shorter. So, the ratio of the man's height to his shadow (*a*/*b*) must get larger. This ratio is called the **tangent** of the angle. Likewise, the ratio of the man's height to the length of the sloping side of the triangle is called the **sine** of the angle, and the ratio of the shadow to the sloping side is the **cosine**. Hipparchus worked out these ratios for every possible angle and compiled his results into a table of numbers. With these tables, he could figure out the height of any right-angled triangle just by knowing the angle and the length of one of the triangle's sides. He'd invented a whole new branch of maths: trigonometry.

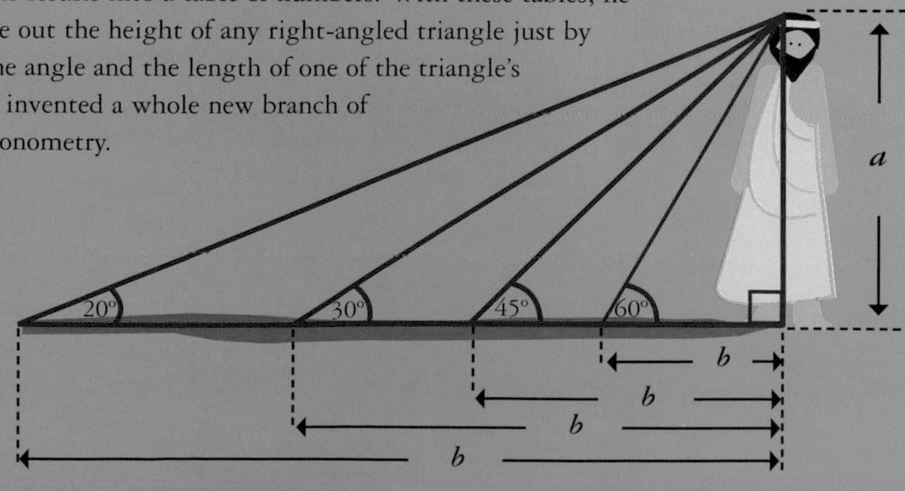

20°
30°
45°
60°
a
b
b
b
b

Measuring the Moon

Hipparchus didn't invent trigonometry just to measure how tall things were. He used it to study the motion of the Sun, Moon, and planets. He used a nifty bit of maths to calculate how far away the Moon is. To do this he made measurements of the Moon's size and position when it was directly overhead and compared them to measurements when the Moon was on the horizon. By drawing a right-angled triangle, he correctly calculated that the Moon's distance from Earth is 30 times Earth's diameter.

Trigonometry has all sorts of uses in the modern world, from calculating the force a pair of pliers can exert to digging tunnels. In 1905, builders used trigonometry to build the Simplon tunnel through the Alps mountains. They dug from both ends, but they couldn't set a course by looking towards the other end because the mountain was in the way. So instead they created an imaginary triangle linking the ends of the tunnel with another point from which the ends of the tunnel could be seen. They worked out the angles and started digging. Eventually the two teams met in the middle – only 10 cm (4 in) apart.

Tricky trig

Trigonometry may sound like a scary branch of maths but it's actually quite simple. It's just a way of figuring out the dimensions of a right-angled triangle from a few bits of information. The only really tricky thing about trig is learning the jargon – the technical words like sine, cosine, and tangent. These words tell you the ratio of any two sides. Fortunately, there's an easy way of remembering them all ...

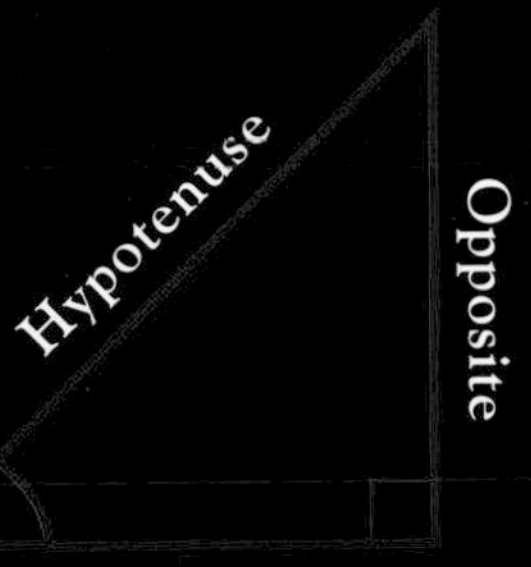

This is the bit you need to learn at school:

Sine = **O**pposite / **H**ypotenuse
Cosine = **A**djacent / **H**ypotenuse
Tangent = **O**pposite / **A**djacent

So how can you remember it? Easy. Just learn the word SOHCAHTOA or try and remember this phrase: *Some Old Happy Cats Are Having Trips On Aircraft.* Or this one: *Some Old Hags Can't Always Hide Their Old Age.*

A round *world*

Until about 3000 years ago, people had little idea what size or shape the world was. In Mesopotamia (now Iraq), people thought the world was a flat disc floating in a giant ocean. Elsewhere in the Middle East, other people thought it was a huge dome with holes through which the Sun rose and set. It wasn't until sailors began exploring the seas that people began to realize the amazing truth: that the world is curved round to form an enormous sphere.

Phoenician sailors

The first people to realize the world is round were probably the Phoenicians, who lived around **3000 years ago** in the land we now know as Lebanon. Unlike the deserts of Arabia and North Africa, Phoenicia was green, with mountains and forests. The Phoenicians used their timber to build mighty ships that could sail hundreds of miles across the Mediterranean and beyond. They travelled south around Africa to buy slaves, and north around Europe to the Isles of Scilly in Britain to buy bronze.

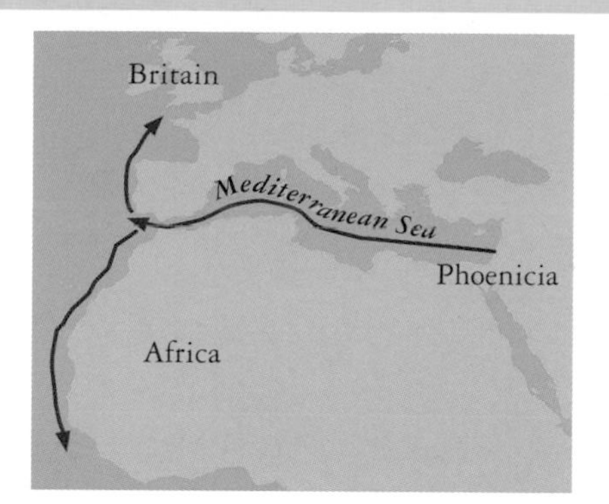

3

Land ahoy!

Phoenician sailors discovered something curious about the way land appears as they approached the shore. They didn't simply see an island grow larger – they saw it come into view from top down. The tips of mountains became visible first. Lower slopes and hills came into view next, and finally the shore appeared. Likewise, merchants waiting in port would see the tip of a ship's mast first, then the sail, and finally the hull. The same thing happened wherever the Phoenicians sailed, and it proved that the sea's surface wasn't flat but was curved.

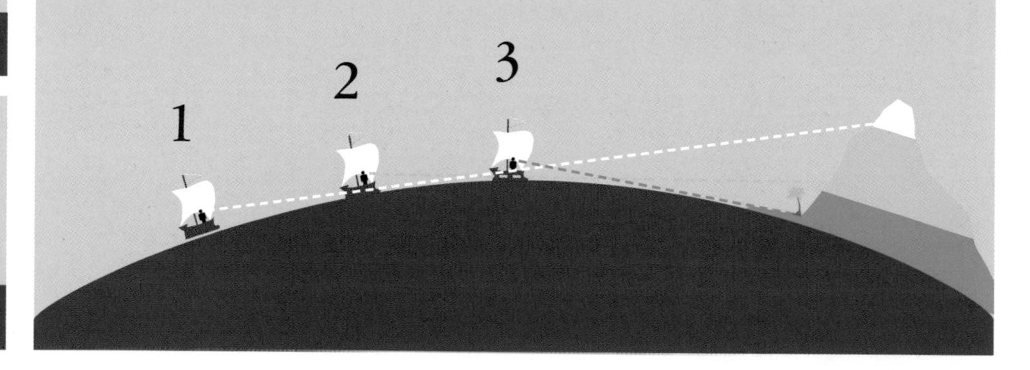

A sailor's view

The distance a sailor can see depends on how high his eyes are. A small rise above sea level can make a big difference – even standing on someone's shoulders allows you to see a mile further. To see twice as far, you need to be four times as high. For the best view, sailors used to climb to the top of the mast.

HEIGHT ABOVE THE WATER	DISTANCE A SAILOR CAN SEE
1.5 m (5 ft)	5 km (3 miles)
3 m (10 ft)	7 km (4.25 miles)
6 m (20 ft)	10 km (6 miles)
12 m (40 ft)	14 km (8.5 miles)
18 m (60 ft)	17 km (10.5 miles)
30 m (100 ft)	22 km (13.5 miles)

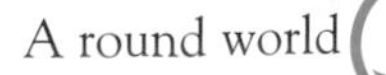

SAILING BY THE SUN The Phoenicians had no compasses to guide them, but they could sail a great distance without getting lost by following the coast. On their long voyages to Britain and Africa they discovered that the midday Sun is lower in the north, making noon shadows longer, but higher in the south, making noon shadows shorter. The reason for the difference was Earth's round shape, which caused the noon Sun to hit different places at different angles. The Phoenicians realized they could use the height of the noon Sun and the length of shadows as a rough measure of how far north or south they'd ventured.

SAILING BY THE STARS Experienced stargazers know that although most stars circle through the sky during the night, one is in the middle of that vast wheel and so always keeps still – the Pole Star. The Pole Star is always due north and has been used by sailors as a kind of compass for many centuries. When the Phoenicians travelled south around Africa, they'd have noticed the Pole Star gradually rising as they travelled north and sinking towards the horizon as they went south. The height of the star was an even better measure of how far north or south they'd travelled than the Sun was, because, unlike the Sun, the Pole Star doesn't move around the sky with the passing hours or changing seasons.

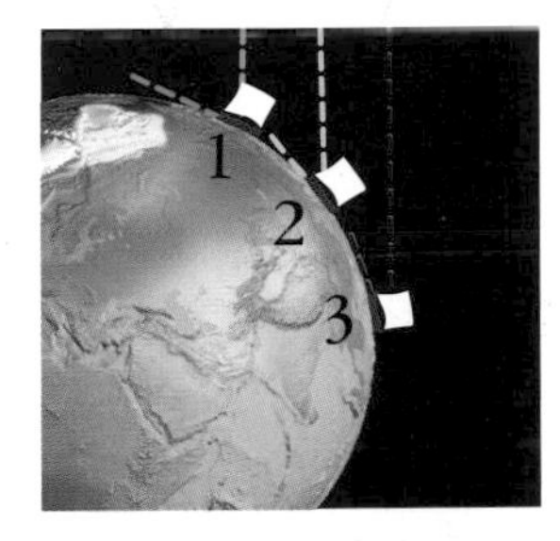

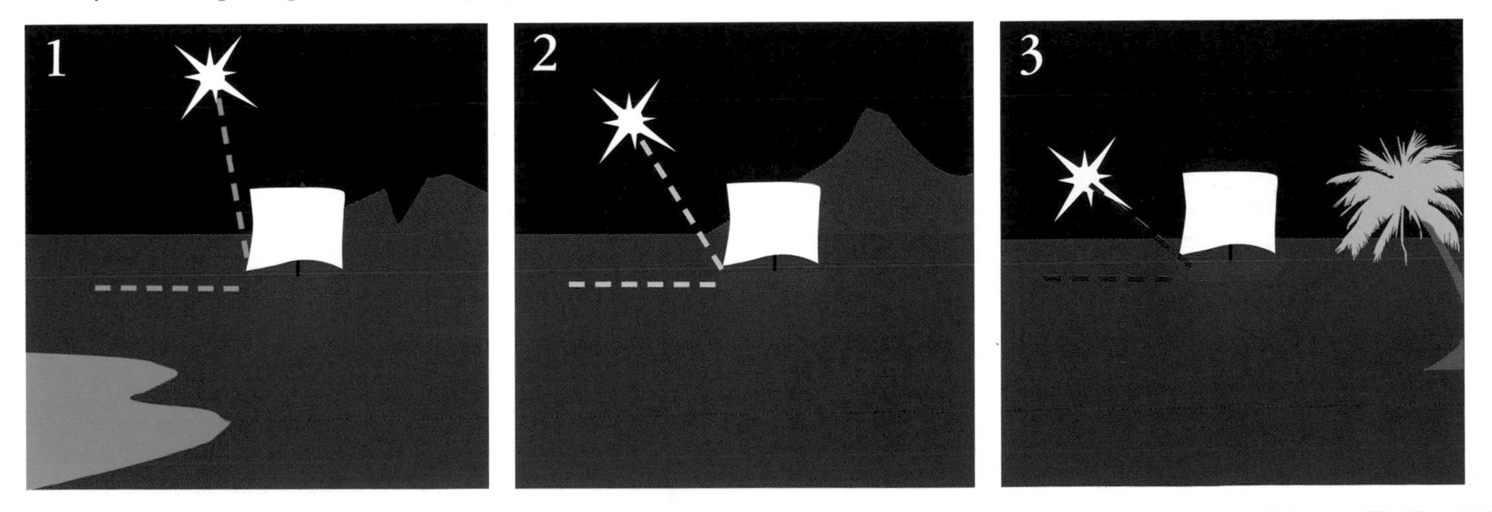

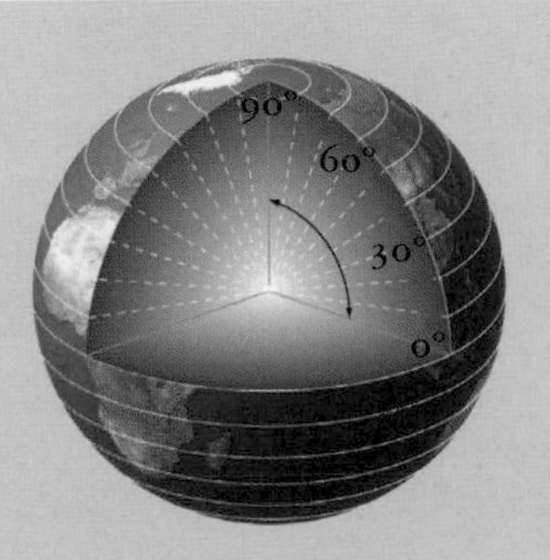

Latitude

The Sun and stars had helped the people of the ancient world to measure the year. Now the Phoenicians found they could also use them to figure out where they were on Earth's curved surface. They'd discovered a rough way to measure what we now call **latitude**. Latitude is the angle between the equator (Earth's middle) and any point on the planet, and it tells you how far up or down (north or south) you are on a map. Later sailors were to discover that you can measure it with great accuracy simply by reading the angle of the Pole Star. But as we'll find out later, it was many centuries before sailors figured out how to measure how far east or west they'd moved and so determine their **longitude**.

Measuring the world

Like the Phoenicians before them, the Greeks knew the world was round. But one brilliant Greek man went a step further: **he calculated our planet's size** – with amazing accuracy.

One of the cleverest

mathematicians of ancient Greece was a man called **Eratosthenes**. Eratosthenes lived not in Greece but in the Egyptian city of Alexandria, which in 240 BCE was the capital of the Greek empire. A brilliant writer and teacher, he was put in charge of the great library in Alexandria, where the Greeks kept all their precious knowledge written down on scrolls of paper.

One day,

Eratosthenes came across a story that fascinated him. He read about a very unusual well in the far south of Egypt, in the town of Syene. At only one moment each year – at noon on midsummer's day – a beam of sunlight shone right down the well and struck the water deep at the bottom, which then reflected the dazzling beam of sunlight back up like a mirror. Eratosthenes realized the Sun must be exactly overhead at this moment, so that its rays hit the ground perfectly vertically, casting almost no shadows.

But the Sun

didn't do the same thing in Alexandria in the north of Egypt. On midsummer's day in Alexandria, sunlight struck the ground at a slight angle, casting small shadows. Eratosthenes found a tall pillar and measured its height and shadow. He drew a triangle, and from this he worked out the sunlight's angle. It was 7.2° off vertical.

7.2°

The Greeks knew that sunbeams always travel in parallel rays, so the difference between the angle of the rays in Alexandria and Syene had to be due to the curvature of the Earth. The Phoenicians had already discovered Earth was round, but now Eratosthenes had enough information to work out its size.

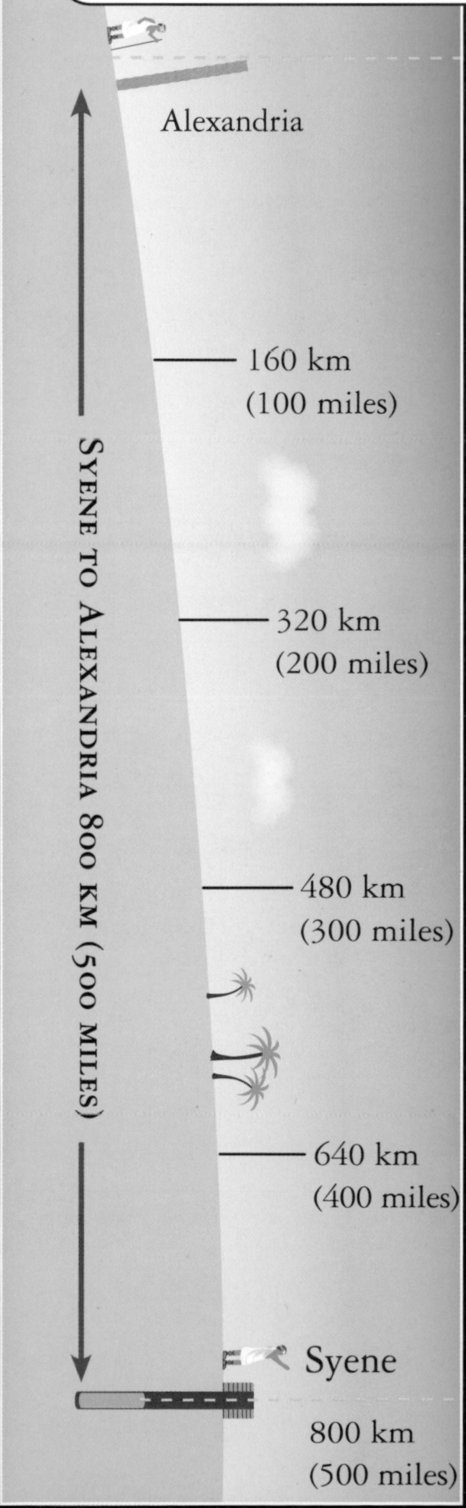

He imagined two straight lines passing down through the well and the pillar and extending all the way to the centre of the Earth, where they would meet. He saw these two lines must also meet at an angle of 7.2°. Since 7.2° is one fiftieth of a circle, all Eratosthenes had to do to calculate the size of Earth's circumference was find out the distance from Alexandria to Syene and multiply it by 50. The answer he got – 40,000 km – was almost exactly right.

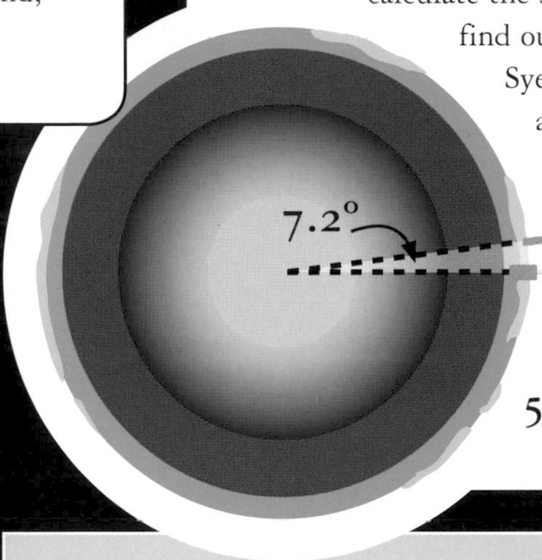

7.2°

7.2°

360° ÷ 7.2° = 50

50 × 800 km = 40,000 km

Alexandria

160 km (100 miles)

320 km (200 miles)

480 km (300 miles)

640 km (400 miles)

Syene

800 km (500 miles)

SYENE TO ALEXANDRIA 800 KM (500 MILES)

Eratosthenes used his discovery to draw a new map of the world. He even put lines of latitude on it, which he worked out, ingeniously, by comparing the lengths of midsummer's and midwinter's day. But no one took him seriously. The problem was that the world was far bigger than they'd ever imagined. Eratosthenes said there must be vast continents or oceans that lay undiscovered, but people found this hard to believe. He also thought there was more ocean than land, with the seas joined to form one great interconnected body of water, and he was right again.

BEHOLD MY NEW MAP OF THE WORLD!

YEAH, OK, IF YOU SAY SO! HA HA !

Eratosthenes didn't live to receive the recognition he deserved. He died at the age of 80 – by then blind and miserable – by starving himself to death. It wasn't until about 1700 years later that he was finally proved right. In the meantime, hundreds of sailors perished at sea by following maps that got the size of the world completely wrong.

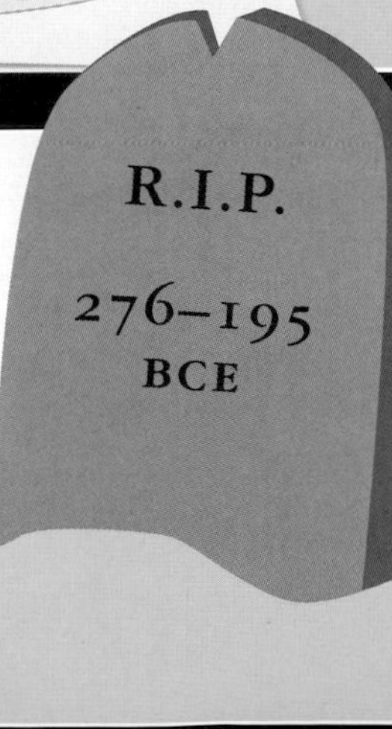

R.I.P.

276–195 BCE

WHY PI?

Once Eratosthenes had figured out Earth's circumference, he could have worked out its diameter (width) too. But to do that he would have needed a special number that has fascinated people since well before the age of ancient Greece: pi. Pi (pronounced "pie") is the ratio of a circle's circumference to its diameter and we use the Greek letter π to write it.

What exactly is pi?

As we now know, π is about 3.14. We can't say what it is exactly because π's decimal places stretch on forever with no pattern. It's impossible to define π as a ratio between two whole numbers, so we call π an *irrational number*. And there's no straightforward equation that can be used to calculate π, so we also call it a *transcendental number*. All of which not only makes π impossible to calculate precisely but also downright weird.

Squaring the circle

The ancient Greeks loved to solve geometry puzzles using only a ruler and a pair of compasses. For instance, they figured out how to make a hexagon out of a circle...

1 The compasses are used to draw a circle and then mark it with arcs of equal size.

2 Straight lines are drawn to join the cross points.

3 Bingo: a perfect hexagon!

...but they thought up one geometry puzzle that well and truly stumped them.

The challenge was to draw a circle and then use it to create a square with the same area. This was called "squaring the circle". The Greeks never solved it, and

The search for pi

The fact that π is impossible to calculate exactly didn't stop people trying. The problem was to accurately measure *the distance around a circle* (measuring the diameter was the easy bit). The Egyptians had a try. They drew the figure below – a circle with a hexagon inside. The hexagon is made of six equal-sided triangles. You can see that the distance around the hexagon is 6 sides and across it is 2. So the ratio of the hexagon's circumference to its diameter is 3. Now the distance across the circle is 2 sides, but the distance around it is clearly more than six sides, so π must be more than 3. The Egyptians made a pretty good guess at $16^2/9^2$, which is 256/81 or 3.16.

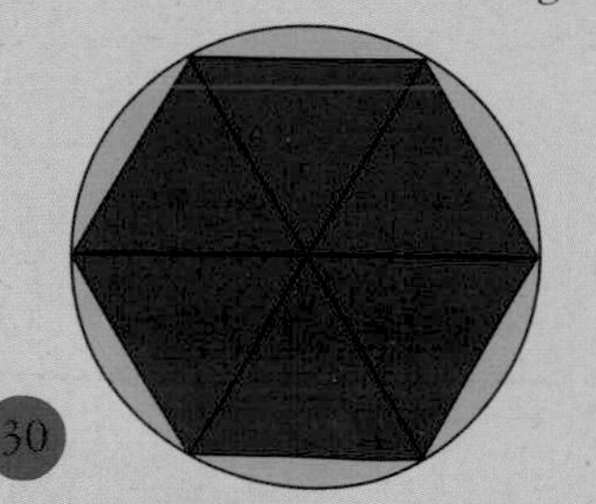

Getting closer...

Around 250 BCE, the Greek mathematician Archimedes got even closer to π by sandwiching circles between other shapes. This ingenious technique allowed him to get closer and closer to measuring the distance around a circle. In the example to the right, the distance around the circle must be somewhere between the distance around the two squares. Archimedes realized he could get an increasingly precise answer by adding ever more sides to his shapes.

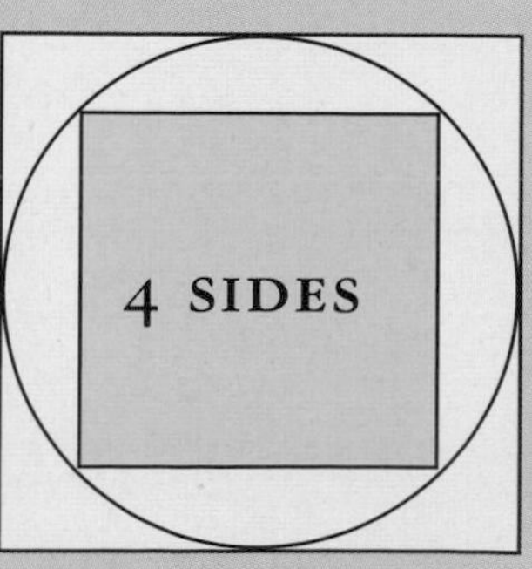

Why pi?

People of the ancient world knew π was important, but they could never have guessed how useful it would become. Today, scientists and engineers use π for an amazing range of calculations involving circles and curves, from planning the routes of airliners to analysing sound waves.

Circumference

Diameter

Radius

Pi is the circumference of a circle divided by the diameter

53589793238462643383279502884197169399375105820974944592307816

today we know why. Solving the puzzle involves using the square root of π (√π) to construct the square, but square roots are impossible to calculate for transcendental numbers.

Building with circles

Circles were not just a mathematical curiosity to the Greeks. They were used in building semicircular theatres, the curving shape not only giving each member of the audience a good view but amplifying sound. Though impressive, Greek theatres were simple structures built into natural, bowl-shaped hollows. But the next civilization in our story was to use circles and curves to create some of the most amazing buildings in the world.

The Theatre of Dionysos in Athens, Greece.

... and closer

He tried 6-sided shapes (hexagons) and got a bit closer. The more sides he added, the nearer their edges got to the circle and the better his answer became. He carried on until his shapes had 96 sides and looked almost indistinguishable from circles. The 96-sided shapes told him that π was between 3.1428 and 3.1408 – a brilliant achievement. It remained the most accurate value for π until Chinese mathematicians improved on it just over 500 years later.

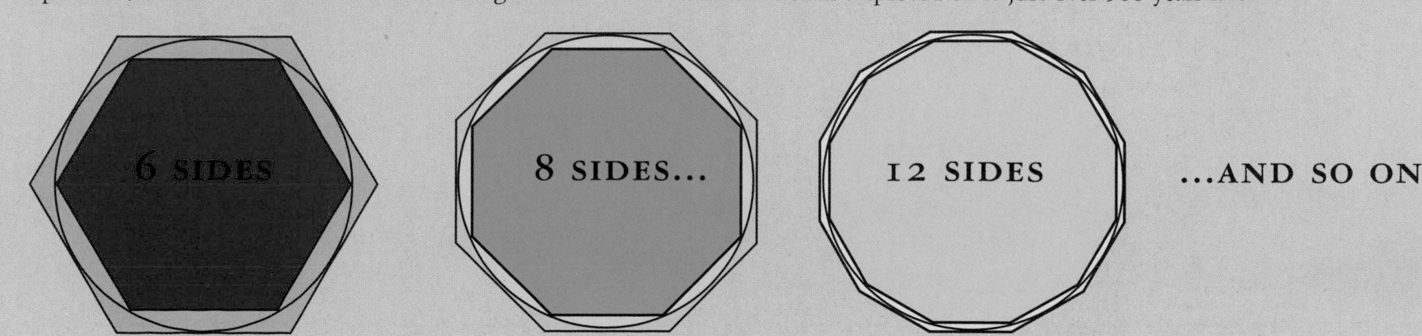

BUILDING *Rome*

The Greeks were great architects, but their buildings were mostly simple rectangles with roofs supported by lots of columns. However, the Romans, who conquered Greece and took over the Mediterranean, had better ideas. Their mighty empire stretched across Europe and North Africa. For the Romans, spectacular buildings were the way to dominate and impress their subjects. Many of these structures can still be seen today – proof of the Romans' engineering skills.

The ARCH

The Romans were quick to discover the advantages of semi-circular arches, which are superb at carrying weight yet require little stone to build. Each stone in an arch is held firmly in place by the weight of the stones above it. If a load is placed on top, its weight is spread evenly through the arch and down through the pillars, making arches amazingly sturdy. An arch is such a strong structure that you can square it off at the top and then build another one above it.

The SPHERE

The Romans' expertise with arches reached new heights in the Pantheon, a temple designed around a sphere. It still stands in Rome today. The heavy semispherical roof was made not of stone but concrete (which the Romans invented) and had a central hole that acted as a sundial. If the curving shape were continued to the ground, it would form a perfect sphere touching the floor at one point in the centre.

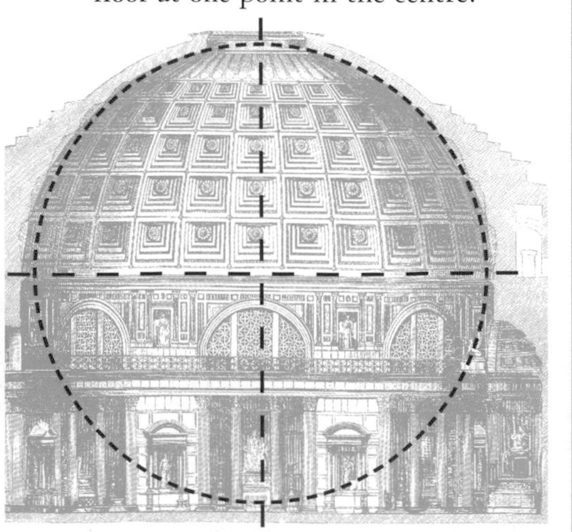

THE COLOSSEUM

The Romans were not only good at using circles and spheres in their designs – they also used ellipses. An ellipse is a perfect oval shape that is longer than it is wide. The colosseum was a huge elliptical building in Rome and was the entertainment palace of its day. Built on a massive scale, it was where you went to watch gladiators, prisoners and wild animals fight to the death.

Easy ellipses It's a mystery how the Romans created ellipses, but they might have used a method like this. Tie a piece of string into a loop and place it round two pins. Place a pen in the loop and pull the string tight. Move the pen around the pins, keeping the string taut and making sure it doesn't slip. Hey presto – you have an ellipse!

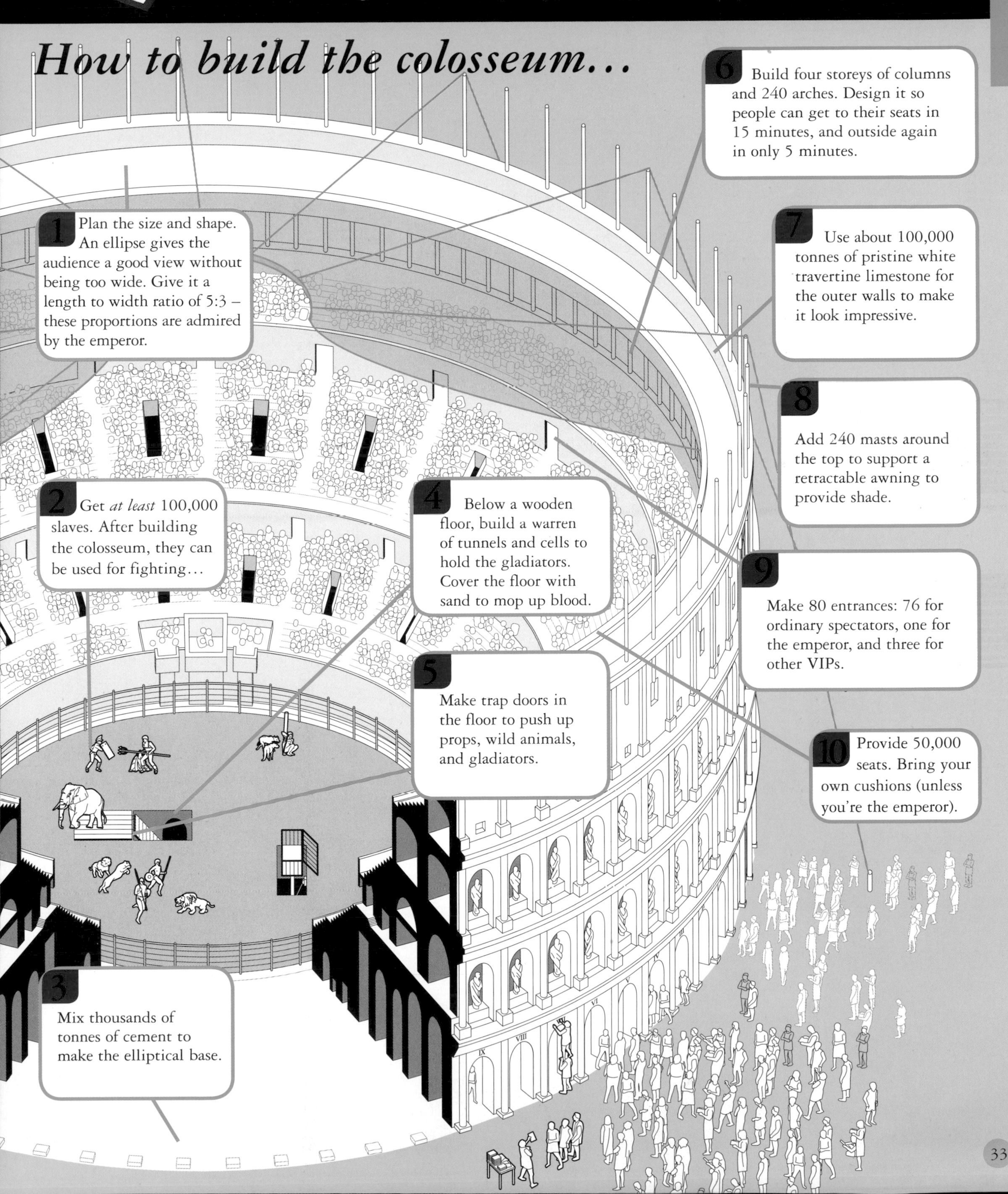

The art of building aqueducts

Strong, useful, beautiful! That's what the Romans' aspired to in their buildings. Their mighty aqueducts transported water into cities from up to 100 km (60 miles) away, with an almost invisibly gentle slope all the way along. They were an astounding feat of engineering, proving that the Romans had mastered the art of measuring. Aqueducts provided such an abundance of water that Roman cities were awash with bubbling fountains and luxurious bathhouses.

The city

Roman cities needed lots of water. In fact, the Romans used almost three times the water we use per person today. The bathhouses alone were enormous. The Baths of Caracalla in Rome, for example, were bigger than a football stadium.

Arcade

A beautiful arcade was the perfect way to keep the water elevated so it could flow down into the city. Arcades were bridges built with a series of arches. They used less materials than a wall and allowed people to pass beneath them easily. Many have become famous monuments of the Roman Empire.

Wall

If the water had to be just a few metres above the ground, the Romans built a wall with a channel on the top. If the height needed was more than 1.5 m (5 ft), they built an arcade instead.

The gradient was amazingly gentle – the water dropped as

Pont du Gard, France

This awesome structure was part of a nearly 50-km (31-mile) long aqueduct that carried water to the Roman city of Nemausus (Nîmes). The aqueduct delivered 20,000 cubic metres (5 million gallons) of water a day.

The top man

One of the great architects of Roman times was a man called Vitruvius. A lot of what we know about ancient Roman architecture comes from his book *De Architectura*. He wrote about how to plan and build aqueducts, saying that they should drop no more than 1.3 cm (0.5 in) in height for every 30 metres (100 feet) in length so that the water would flow gently. How the Romans achieved this amazing feat with their simple measuring tools remains a mystery.

LAKE

Tunnel

If there was a mountain in the way, a tunnel was made. Vertical shafts dug down from the surface made the job easier. These were left open after the tunnel was finished so that slaves could be sent down to clear limescale deposits left by the water, which might otherwise block the tunnel.

Siphon

Lead pipes called siphons were sometimes used to carry water across valleys. Provided the water level at the start of the siphon was higher than at the end, the water would flow up the downstream part of the siphon without needing to be pumped.

Trench

Four out of five miles of Roman aqueducts were covered trenches. They were sometimes lined to stop the water seeping out.

little as 30 cm (1 foot) over a distance of 1 km (0.6 miles)

Roman roads were ***incredibly straight*** and stretched for thousands of miles, allowing Roman armies to get around ***FAST***. But how did the Romans get them so straight?

Roman roads were built dead straight by line of sight, from hilltop to hilltop, only bending at river crossings. The Romans planned the routes with the help of a tool called a *groma*. This consisted of an upright pole with two sticks on top, arranged at right angles to form a cross. From the ends of the cross hung weights on string (plumb lines). The plumb lines ensured that marker poles were upright and always perfectly in line. The groma also helped builders to get the slope of aqueducts right.

Why MEASURE *any body?*

The world's first measuring instrument was the human body. Before people invented rulers or other devices to measure the size of something, they simply compared it to their body. Even today we use the names of some body parts as units of size or distance.

Throughout history people have used their fingers to count and their hands, arms, and legs to measure. The largest body measurement was a person's height; the smallest was a hair's-breadth.

THE YARD The English King Edward the First used the distance from his nose to his outstretched fingers to represent a yard. Today most people measure not in yards but in metres, which are slightly longer, so tailors turn their head to the side to add the extra 86 mm (3.4 inches).

1 yard

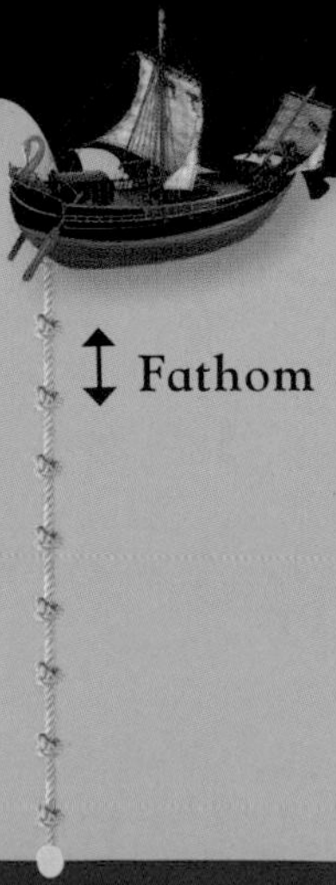

THE FATHOM Sailors measured out rope two yards at a time by stretching it from hand to hand. They called this distance a fathom. By tying a weight on the end of a rope knotted in fathoms, they could measure the depth of shallow waters and check whether a big ship might run aground.

Fathom

THE CUBIT is a very ancient body measurement that was based on the length of a man's forearm from elbow to fingertips – 18 inches or 457 mm. In the bible, Noah's Ark was said to be an enormous 300 cubits long (137 m or 450 ft) by 30 cubits wide (14 m or 45 ft) tall. In fact, no ship so big was built until the year 1858.

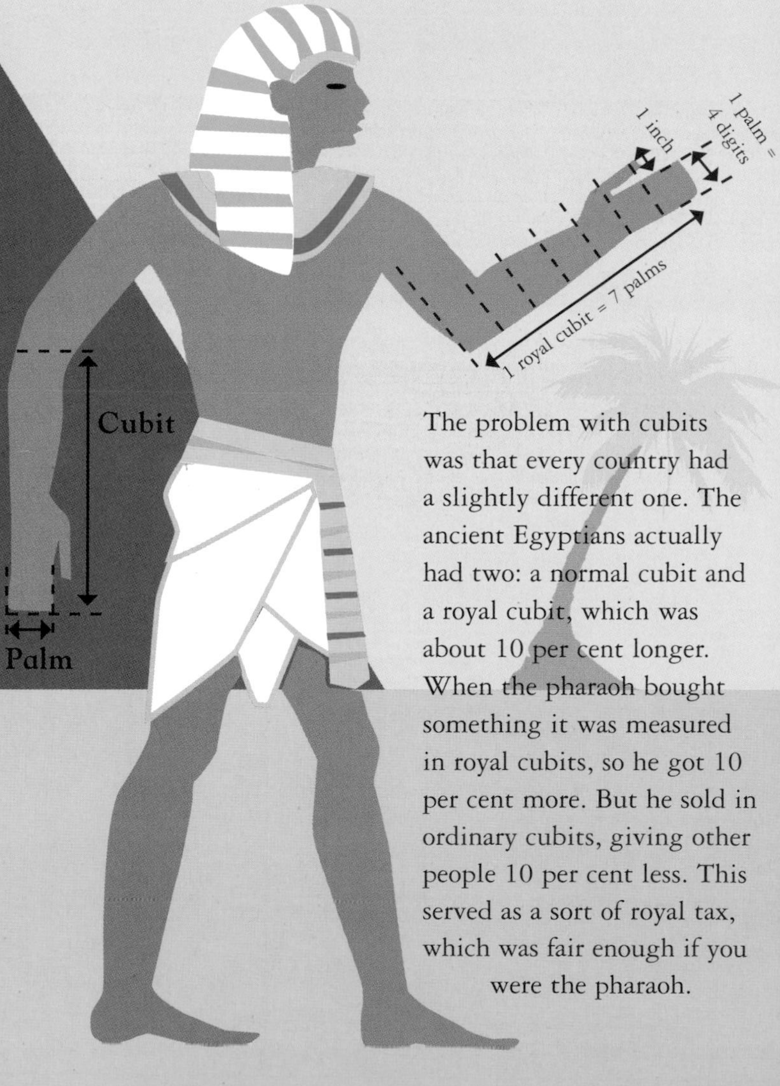

1 palm = 4 digits

1 inch

1 royal cubit = 7 palms

Cubit

Palm

The problem with cubits was that every country had a slightly different one. The ancient Egyptians actually had two: a normal cubit and a royal cubit, which was about 10 per cent longer. When the pharaoh bought something it was measured in royal cubits, so he got 10 per cent more. But he sold in ordinary cubits, giving other people 10 per cent less. This served as a sort of royal tax, which was fair enough if you were the pharaoh.

Is the following statement TRUE OR FALSE:

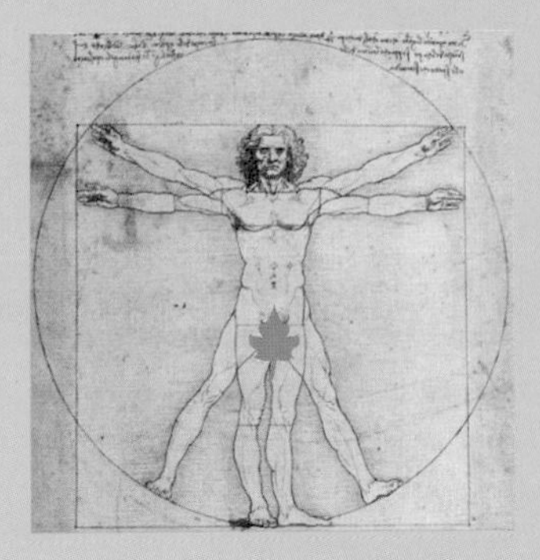

In 1490, the Italian artist Leonardo da Vinci drew a famous painting inspired by the Roman architect Vitruvius and ancient Roman measurements. Called Vitruvian Man, it features a nude man with outstretched arms and legs inscribed in a circle and square. The cubit, foot, palm, and pace are all accurately shown and in perfect proportion.

The Vitruvian Man painting shows that a person's armspan (1 fathom) is about the same as their height (1 stature). Try it for yourself. Stand against a wall and mark your height. Then see if your fingers can stretch the whole way.

THE ROMANS measured long distances by pacing. One pace, or *passus*, was about 1.6 m (5 ft). If you lay a tape measure on the floor you'll see this is actually two steps. A professional pacer measured the distance between towns by counting only his right or left foot. One thousand paces – *mille passuum* – became known as a mile, a unit we still use today, though the Roman mile was about 129 m (423 ft) short of the "statute mile" we have now.

Today, some soldiers count in eights as they march by chanting these words: "Left, left, I had a good home and I left!"

The Romans also used the foot (*pes*). The Roman foot was one fifth of a pace and one sixth of a man's height, or about 29.5 cm (11.6 in). But it was longer than the average human foot, just as the modern imperial foot is. Perhaps the Roman foot included the tough leather sandal that Romans wore.

FINGERS AND THUMBS don't make very reliable measuring devices as everyone's are different. That's why "rule of thumb" means approximate. The inch may be based on the width of a man's thumb across the middle knuckle. In many languages (including French, Spanish, Italian, Swedish, Portuguese, and Dutch), the word for inch is also the word for thumb.

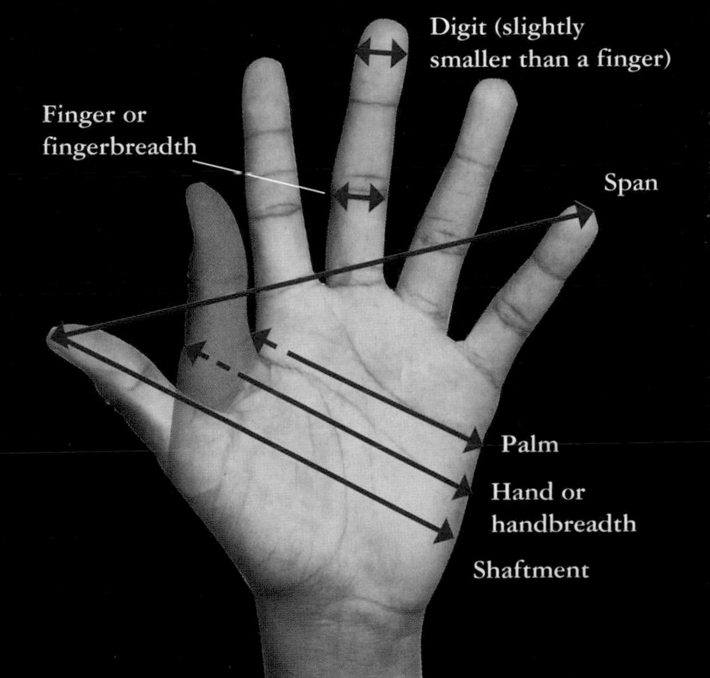

THE HAND is the width of a hand with the thumb closed. This old-fashioned measure is now only used to measure the height of horses from foot to shoulder. An adult animal less than 14.2 hands is called a pony. Anything taller is a horse. The world's smallest "horse", called Thumbelina, is just 4 hands tall.

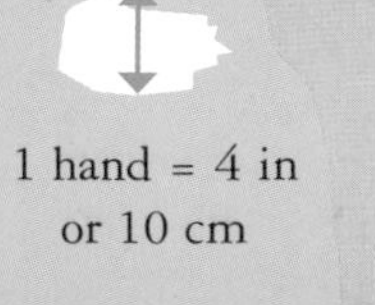

1 hand = 4 in or 10 cm

"The vast majority of people have more than the average number of legs."

Find out the answer at the back of the book!

Night

We can all tell roughly what time of day it is just by looking to see how bright it is outside. In the past, people also found

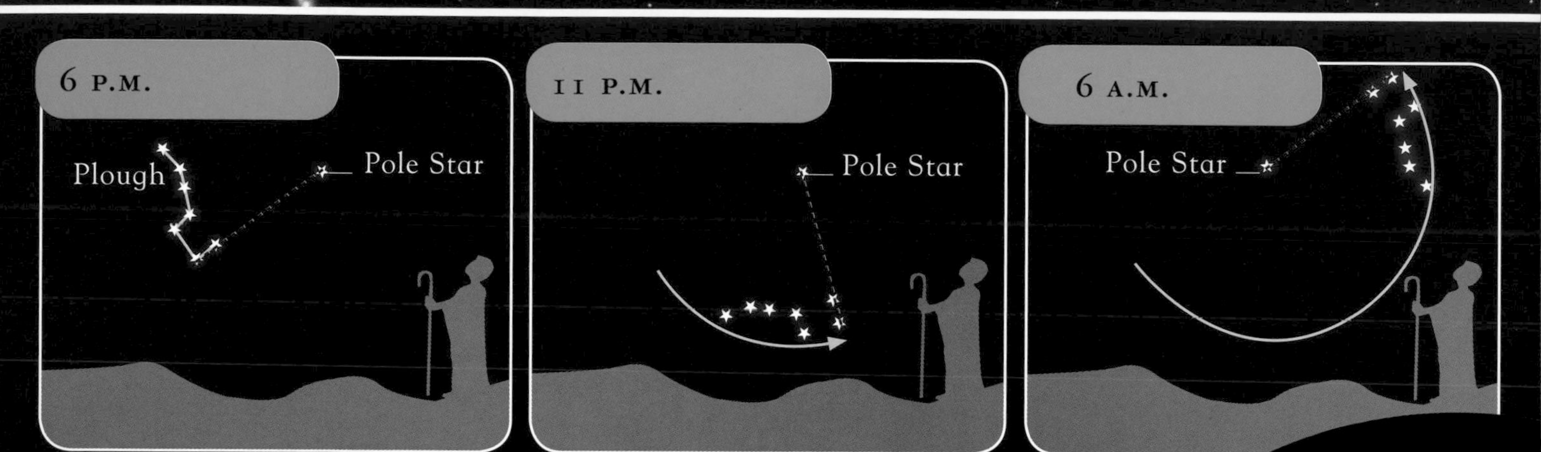

Star time

As we saw earlier, the Pole Star is the only star to keep still in the night sky, while other stars circle around it due to Earth's rotation. Early people found they could measure time by watching the constellations move around the Pole Star like the hour hand of a clock. To find the Pole Star, look for the Plough (Big Dipper). The Pole Star lines up with the Plough's last two stars.

Star clock

During the Middle Ages, people used a device called a nocturnal to read the time from the stars. This had a rotating arm that was lined up with the last two stars of the Plough (the "pointers"). Find out how to make your own star clock overleaf.

There are 24 hours in one day

Fire time

Once people discovered fire, they had something to keep them warm, to cook with, to scare off predators, and to light their homes. They made lamps by pouring oil into small containers with a wick and setting fire to the wick. The oil level in these oil lamps slowly dropped as the hours passed, providing a measure of time. Later on, candles marked with hour-lines did much the same job.

SEASHELL OIL LAMP

We have days and nights because the Earth rotates.

CANDLE CLOCK

Who decided to divide the day into two sets of 12 hours?

AND *day*

ways to use the stars, fire, water, and shadows to measure the time of day or night more precisely.

Egyptian *merkhet*

The ancient Egyptians created a timekeeping device called a *merkhet*. Two people were needed. One looked through a V-shaped notch in a stick towards the other in order to take an accurate sighting of a star or the Sun. At night the *merkhet* worked like a star clock, but it could also be used with the Pole Star to make a north–south line in the ground. When the Sun cast a shadow along the line, it was noon.

Earth's axis

Sands of time

The hourglass was first used about 700 years ago. Sand slowly trickled through a neck between two glass bulbs. When the top bulb was empty, a measured amount of time had passed. This wasn't necessarily an hour – it could be anything from five seconds to a year! Small 3-minute hourglasses called egg timers are still used today for timing boiled eggs.

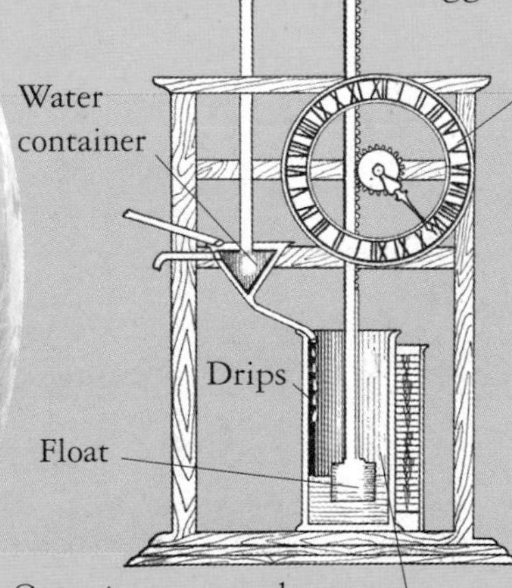

Water clock

The water clock worked in a similar way to an hourglass but relied on water dripping through a tiny opening rather than trickling sand. The ancient Greeks used a water clock called a *clepsydra* (left), which means "water thief". In early water clocks the container collecting the drips was marked with hour lines, but later versions were more sophisticated and had a clock face with an hour hand turned round by a rising float.

Shadow clocks and sundials

The ancient Egyptians made a simple but effective shadow clock (left). It was placed on an east–west line, with the raised bar facing the rising Sun in the morning and turned around to face west in the afternoon. Sundials (right) have been made through the ages, in all kinds of different designs. Here, the shadow cast by the central metal "gnomon" falls on a circular clock face.

Shadow

EGYPTIAN SHADOW CLOCK

SUNDIAL

Make a *sundial*

With a few bits of wood, a pencil, and some plasticine, you can make a sundial that will tell you the time whenever the Sun's shining. (Note: you'll need help from an adult to cut the wood.)

The upright part of a sundial that casts a shadow is called a "gnomon". The world's largest sundial is at Sundial Bridge on the Sacramento River in California, USA. Its gnomon is an amazing 66 m (217 ft) tall. Sundial Bridge only tells the time perfectly on 21 June.

Setting the gnomon

For a sundial to work well, the gnomon must line up with Earth's axis. To ensure it does, set the gnomon at an angle equal to your latitude. Use an atlas or the internet to find the latitude where you live. New York, for instance, has a latitude of 40°.

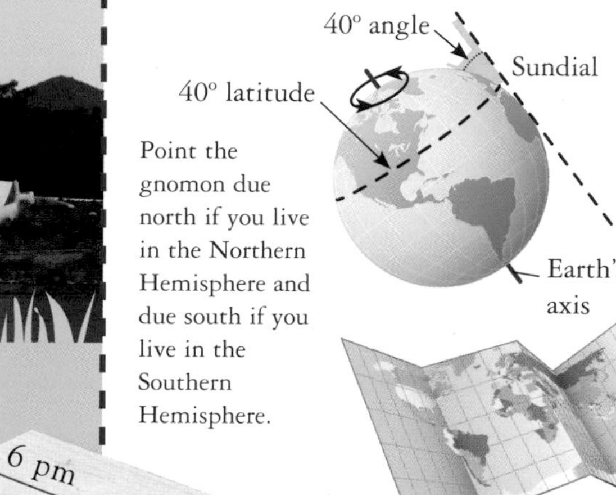

Point the gnomon due north if you live in the Northern Hemisphere and due south if you live in the Southern Hemisphere.

Step 1

Use a pencil to **draw a straight line** across a wooden board from one side to the other. Mark the centre of the line with a dot. **Draw lines 15° apart** coming out from this point.* Go over the lines afterwards with a pen. **Write in the times** as shown on the finished board (right).

Make sure your sundial is pointing the right way!

The angle here needs to match your latitude.

Step 2

Look in an atlas to find your latitude. When you've worked out your latitude, ask an adult to cut a piece of wood into a triangle. One corner of the triangle must have an angle equal to your latitude.

Step 3

Use a lump of Plasticine to **stick a pencil on the dot.** Place the triangle under the pencil, ensuring the corner that matches your latitude is on the dot. **Fix the triangle down with glue.** Take your sundial outside and place it on a flat surface. Point your gnomon north if you live in the Northern Hemisphere, south if you live in the Southern Hemisphere.

* Using a 15° angle between the lines will make your sundial only approximately right. For a more accurate sundial you need to customize the angles between the hour lines to suit your latitude. You can do this by using a "sundial shadow angle calculator" on the internet.

Make a *star clock*

Impress your friends with your ability to **tell the time** just by looking at the stars! You'll need to be able to locate the Plough constellation and the Pole Star to use this clock. (Note: it only works in the Northern Hemisphere.)

Hold the current month at the top and turn the orange disc to match the stars.

The Plough is called the Big Dipper in North America. The stars make a shape that looks like a big saucepan.

Step 1

You need access to a photocopier or scanner and printer to make this star clock. Photocopy the bottom half of this page at double size or scan and print it at double size. It's best to use slightly thicker paper than normal in order to make your star clock sturdy. Cut out the discs. Make sure you're careful!

Base disc

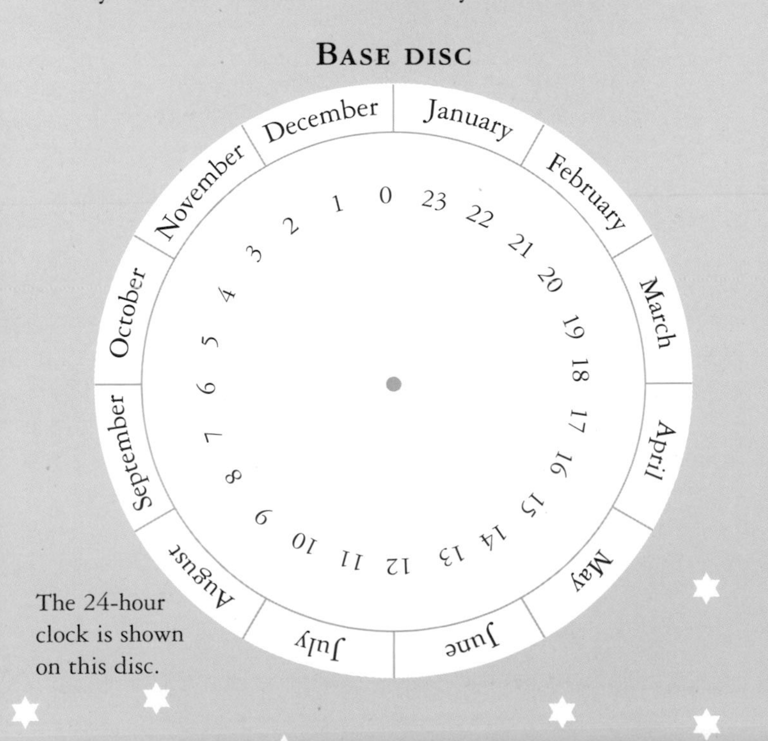

The 24-hour clock is shown on this disc.

Top disc

The time is

Pole Star

The Plough

Cassiopeia

Cassiopeia is opposite the Plough. It's easy to see because its stars form a W-shape.

Step 2

Glue the base disc onto card to make it stronger. Place the top disc on the base disc, punch a hole through both discs and fix the discs together with a paper fastener.

How to use your star clock

Face north and hold your star clock vertically in front of you. Turn the clock so the correct month is at the top. Holding the base disc still, rotate the top disc so the constellations match what you can see in the night sky. Read the time from the small opening.

Weighing up

People have always traded valuable goods. In fact, they started trading long before anyone thought of inventing money. As civilizations flourished and farmers produced ever more goods to trade, people needed better ways to measure value. And so they figured out how to weigh things.

In the balance

Trading is easy if you can count the things you're swapping. You could easily agree that one sheep is worth 20 chickens, for instance. But what if you want to trade something you can't count, like flour, butter, or gold? The fairest way to measure how much you've got is to weigh it. Early traders compared weights by holding things in their hands, but later they invented weighing scales that work like seesaws. The ancient Babylonians used special stones as standard weights. These precious stones were polished and carved into animal shapes. Some people still weigh themselves in "stones" to this day.

Egyptians started using weighing scales about 5000 years ago.

I'll pay you 3 silver shekels for that bag of 50 barley shekels.

You want to buy 50 shekels for only 3 shekels?? Get lost!!!

Making money

Babylonian merchants often exchanged barley for other goods. They weighed out barley into small heaps called shekels, with one shekel equal to about 180 grains. It was such a handy way of paying for things that barley shekels became a kind of money. But you needed a lot of barley to buy something valuable, and merchants got tired of hauling heavy sacks of it. So they began carrying small silver weights instead. These silver shekels, which were worth as much as whole sacks of barley, became the world's first coins. Some modern currencies also started life as silver weights. The British pound, for instance, was originally a chunk of silver weighing exactly 1 pound (0.45 kg).

Silver shekels

Go with the grain

Seeds such as barley were especially handy for weighing small, precious things like gemstones, pearls, and gold. The Babylonians used barley for this purpose, but the Greeks used wheat, and the Arabs used the seeds of the carob tree. The carob weight became our modern unit the "carat", which is still used to weigh diamonds and gemstones today. Since carob seeds vary in weight, the modern carat has been standardized as 0.2 grams precisely.

THE ROMANS also used carats to measure the purity of gold. Their gold coins weighed exactly 24 carob seeds each. As a result, 100% pure gold became known as 24-carat gold. When gold is blended with other metals to make an alloy, its purity is reduced. A mixture of 75% gold and 25% silver, for instance, is called 18-carat gold.

A NEW WAY TO WEIGH

The Greek scientist Archimedes took the science of weighing a step further by figuring out how to measure ***density***. A dense object, such as a rock or a lump of metal, has a lot of weight compacted into a small size. The Greek king gave Archimedes a new crown and asked him if he could figure out, without cutting the crown, whether it was 24-carat gold or a cheaper (and less dense) mix of gold and silver. The solution came to Archimedes when he got into his bath and saw the water level rise. He realized he could drop the crown in water and use the rise in the water level to measure its volume. Then he could weigh the crown and divide the weight by the volume to work out if it was as dense as pure gold. It wasn't. The crown was fake and the goldsmith was sentenced to death.

Legend has it that Archimedes leapt out of his bath and ran naked down the street shouting "Eureka!" ("I've found it!") when he solved the king's puzzle.

Meanwhile, another Greek scientist, called Aristotle, was wondering why weight makes things fall to the ground. He thought heavy things have "gravity" and light things, such as steam, have "levity". As we'll see later, the mysterious and invisible force of gravity, which pulls falling objects towards the ground and gives them weight, turned out to be very important indeed.

PUZZLING WEIGHTS

The branch of maths known as algebra is based on the idea of keeping things balanced, with two sides of an equation always equal like the two sides of a set of weighing scales. Imagine a set of weighing scales with 9 balls on one side but 3 balls and 2 cubes on the other. If the scales balance, they form a kind of equation ($2c + 3b = 9b$). Can you work out how many balls equal one cube?

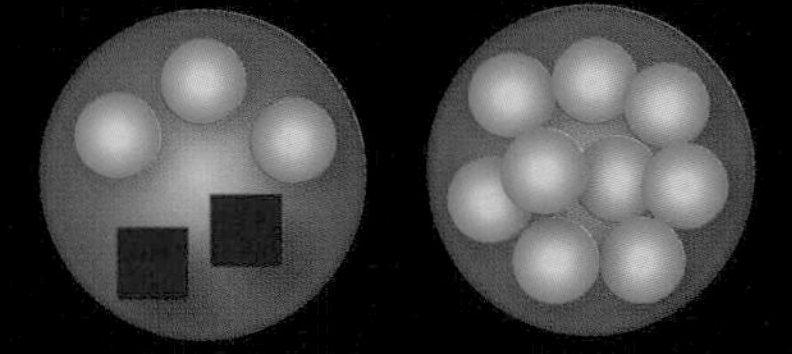

Here's one way of doing it:

1. **Remove 3 balls from each side. The scales still balance.**
2. **So two cubes must equal 6 balls.**
3. **Divide each side by 2.**
4. **So one cube equals 3 balls.**

See if you can solve the next puzzle on your own. You have a set of scales, some fruit, and you find out that the following combinations are balanced:

Can you work out how many plums weigh as much as one orange?

Heavy head puzzle

How could you find out how much your head weighs without chopping it off? Hint: the human body has about the same density as water.

Answers at the back of the book.

The Age of DISCOVERY

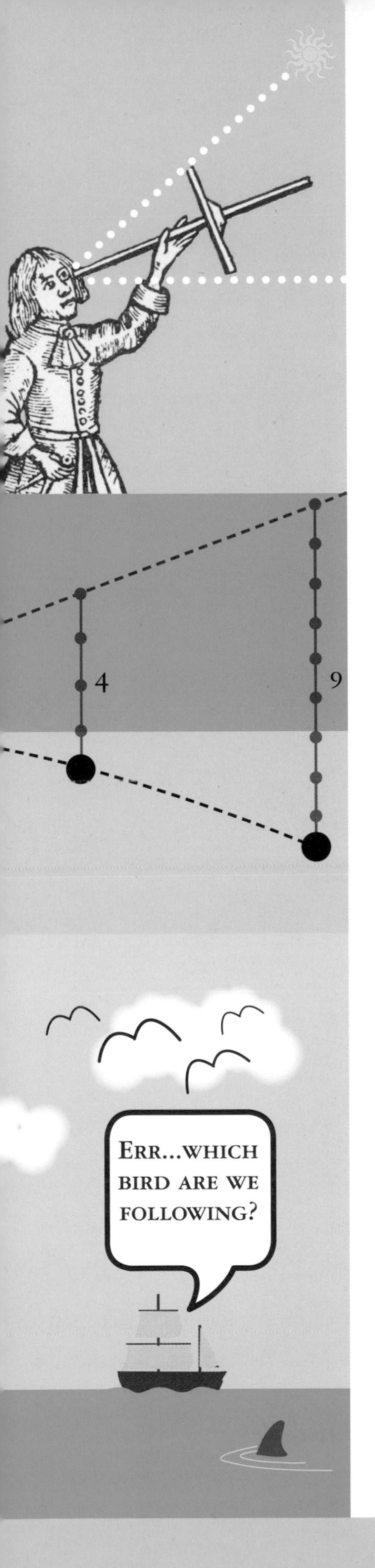

The *mathmagicians* of ancient Egypt, Greece, and Rome built on each other's knowledge, steadily improving their understanding of the world. But when the Roman Empire collapsed around the year 400, Europe went into decline. ***Maths and science went nowhere for nearly 1000 years*** – a period we now call the **Dark Ages.**

In other parts of the world, however, progress continued. In India, Hindu mathematicians invented the INGENIOUS NUMBER SYSTEM we use today. It spread like wildfire through the Arab world, where traders found it made calculations easier.

When the Hindu numbers reached Europe, they helped trigger a revival called the Renaissance. Now science took off as scientists used the new maths to study the mysterious force of gravity and the movement of the planets, and they made some astonishing discoveries. Meanwhile, traders were making ever more daring voyages across the oceans and mapping unknown corners of the globe.

It was a magnificent age of scientific breakthroughs and daring exploration – an *age of discovery*.

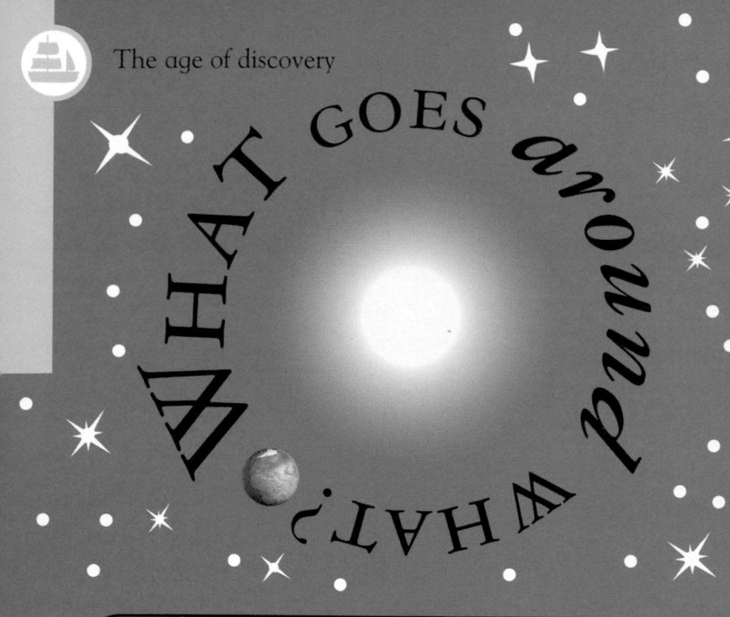

The mathematicians of the ancient world worked out that Earth is round and even measured its size. They watched the Sun and planets travelling across the sky in what looked like great circles, and they naturally thought that Earth was in the middle of everything, with the "heavenly bodies" all circling our world. But it turned out they were hopelessly wrong. Earth wasn't in the middle, and the planets weren't even moving in circles. It took the genius of the two men on this page to uncover the truth.

The Greeks thought Earth was in the middle of the Universe.

Why is a week seven days long?

Ever wondered why a week has seven days and not five or ten or some other number? The reason is that the people of the ancient world counted seven heavenly bodies that moved differently from stars and named the days after them. Here they are in French and English:

Saturday (Samedi)	SATURN
Sunday (Dimanche)	SUN
Monday (Lundi)	MOON
Tuesday (Mardi)	MARS
Wednesday (Mercredi)	MERCURY
Thursday (Jeudi)	JUPITER
Friday (Vendredi)	VENUS

COPERNICUS

In 1507 the Polish astronomer Nicolas Copernicus made what turned out to be an astonishing discovery. He found he could predict the positions of the planets much more easily if he assumed the Sun was in the middle of the Solar System rather than Earth. But this meant that Earth must be flying around the Sun, which was an amazing thought. Even more shocking, it meant that the Sun wasn't really moving across the sky each day, but Earth was spinning round. Copernicus was totally correct, but this clashed with the religious belief that God created the Earth as the centre of things. So as not to upset the Church, Copernicus waited until he was on his deathbed before publishing his theory in a book.

NICOLAS COPERNICUS (1473–1543)

"I demonstrate by means of philosophy that the Earth is round, and is inhabited on all sides; that it is insignificantly small, and is borne through the stars."

JOHANNES KEPLER (1571–1630)

KEPLER

The astronomer Johannes Kepler was born in Germany 28 years after Copernicus died. Copernicus had thought the planets move in circles, but Kepler couldn't quite get circles to fit with detailed observations of the planets' movements. So he kept trying other shapes until he hit upon the answer: the ellipse. Unlike a circle, which has one centre, an ellipse has two (called foci), and it turned out that the Sun is always at one of them. Kepler also found the planets moved faster as they got near to the Sun, as though something was pulling them. As we'll see later in the book, this was to be a vital clue in one of the greatest scientific discoveries ever made.

Kepler's 1st law

Planets don't go around the Sun in circles but in ellipses.

Kepler's 2nd law

Planets move faster when nearer to the Sun. A line drawn from a planet to the Sun will sweep over equal areas in equal times.

HERE'S HOW KEPLER COULD HAVE EXPLAINED HIS THEORY TO PEOPLE WHO AREN'T MATHS GENIUSES...

1. Fasten a ball to a piece of string and pass the string through a cardboard tube.

2. Hold the tube very still with one hand and the string with the other. Pull and release the string to make the ball spin round in circles.

3. While the ball is on its upward curve, pull the string to shorten it a little. Let the string out again on the downward curve. The ball will now move in an ellipse.

All you need for this trick is a ball, a sturdy cardboard tube, and some string (and strong arms).

EXPLANATION

When you pull the string, you'll feel the ball speed up and a stronger force on the string. When a planet gets closer to the Sun, it speeds up in the same way. Kepler thought there must be some kind of "magnetic" pulling force between the Sun and planets, but what was it exactly? The puzzle would later be solved by English scientist Isaac Newton.

GALILEO *the* Great

Around the same time that Kepler lived, an Italian scientist called GALILEO was making scientific breakthroughs that would change the world forever. Galileo was perhaps the world's first true scientist, forming theories and then carrying out careful experiments to see if they worked in practice.

SWINGING TIMES One day in 1581, when 17-year-old Galileo was at church feeling bored, he started watching a lamp swinging in the breeze. Out of curiosity, he timed how long each swing lasted by using his pulse as a stopwatch (watches hadn't been invented). Each swing took exactly the same time, no matter how far the lamp moved. Fascinated by this discovery, Galileo made a pendulum at home and tried lengthening the string to see if it changed the time. It did so, and an amazing mathematical pattern emerged. To double the time of a swing, the string had to be 4 times longer. To triple the time, the string had to be 9 times longer. It was a simple pattern of ***square numbers.***

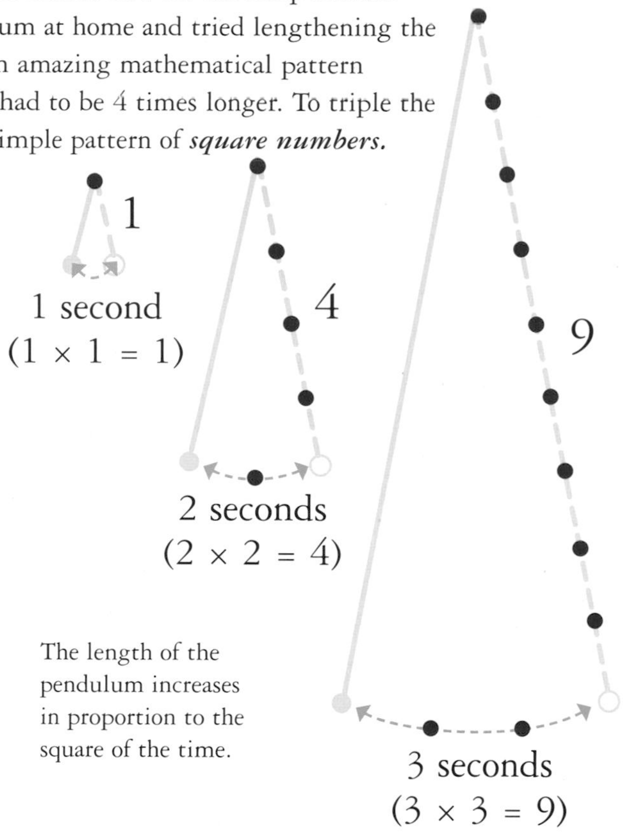

In a grandfather clock, the pendulum keeps the time.

The length of the pendulum increases in proportion to the square of the time.

"Everything in the Universe is understandable, but only if we understand the language it is written in. That is the language of MATHEMATICS."

GALILEO GALILEI (1564–1642)

Tick, tock...

Galileo had discovered that the pendulum could be used to keep almost perfect time. It was far better than any sundial or water clock. When he was older, Galileo even designed a clock regulated by a pendulum, but it was not made until after his death. Clocks and watches regulated by a pendulum or similar mechanism remained the world's best timekeeping devices for the next 300 years.

BEFORE GALILEO'S time, people thought that the heavier an object was, the faster it would fall. But Galileo noticed that the weight of the "bob" at the end of his pendulum made no difference to the time of its swing. He tried dropping different weights from a height (maybe from the Tower of Pisa) at the same time to see if the heavier one landed first. They hit the ground at the same instant, so weight made no difference after all. It was another amazing discovery.

Rolling, rolling, rolling

Galileo noticed his falling weights fell faster and faster – they *accelerated*. This aroused his curiosity yet again. To measure the acceleration, he "diluted" the fall by rolling the balls down a ramp. He had no stopwatch to time them, so he placed harp strings across the ramp so he'd hear bumps as the balls rolled. Then he spaced out the strings so the bumps made a regular beat. He measured the distances to the strings and found the same pattern of square numbers he'd seen with his pendulum.

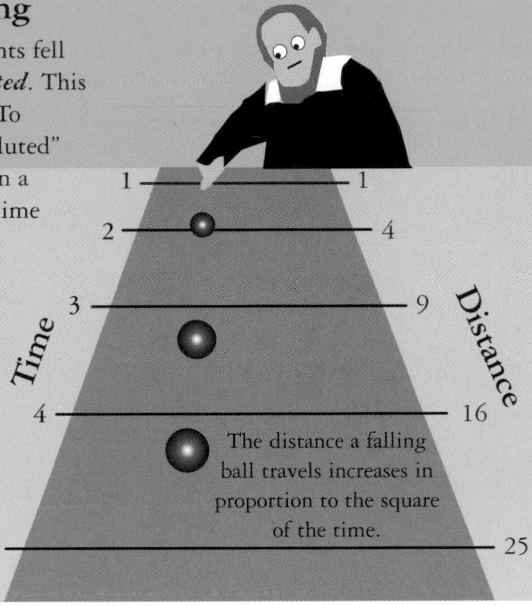

The distance a falling ball travels increases in proportion to the square of the time.

The maths of war Galileo realized he could use his square numbers to figure out the precise curve of a cannonball. A cannonball flies horizontally at a constant speed but falls from the line of fire at an accelerating speed. Just like a ball dropped from a height, it accelerates to Earth in ratio with the square of the time. Thanks to Galileo, soldiers could now calculate a cannonball's path and hit targets out of sight. City walls were no longer a good line of defence and became a thing of the past.

The distance a cannonball falls from a straight line increases in proportion to the square of the time. The result is a curved path called a parabola.

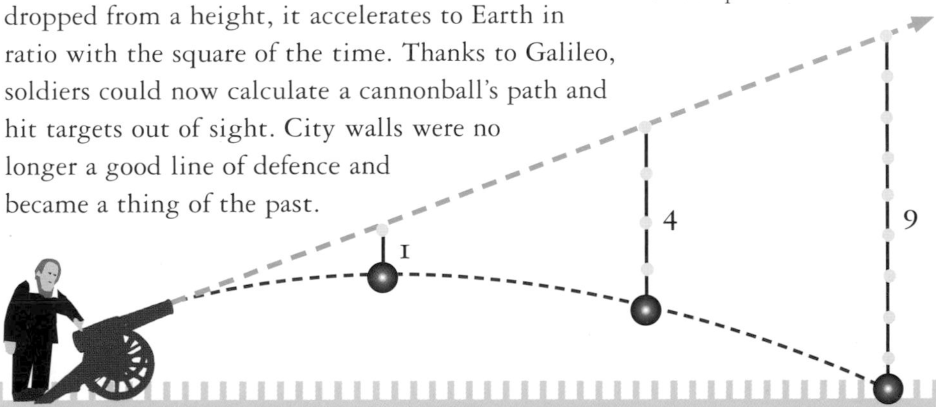

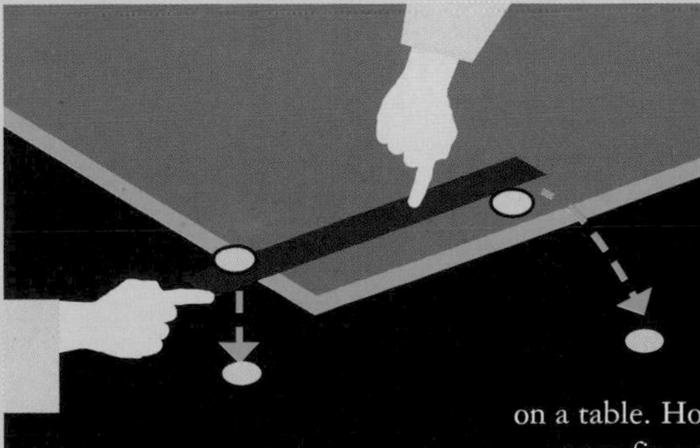

Galileo said a bullet fired sideways would fall at the same rate as a ball dropped to the floor. Prove it with a ruler and two coins. Place one on the ruler and one on a table. Hold the middle of the ruler with your finger to make a pivot, and flick the end. The coins should land at the same time.

The GRAVITY *of the* situation

Galileo had discovered how cannonballs move through the air in a curved path. Kepler had discovered that the planets orbited the Sun in ellipses. But neither man saw the connection. In the year Galileo died (1643), Isaac Newton was born, and he was to put the pieces together and come up with the answer to the puzzle. He called it "GRAVITY".

"If I have seen further it is only by standing on the shoulders of giants."

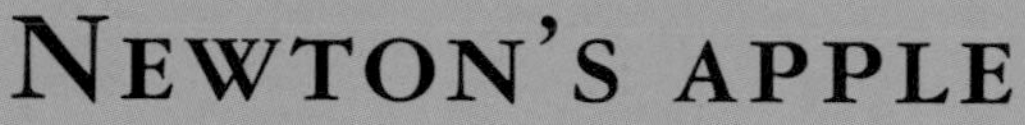

Newton's apple

In 1666, Isaac Newton fled London to avoid the deadly plague that was sweeping through England and went to stay on his mother's farm. Watching an apple fall from a tree, he wondered if the force that pulled the apple to Earth was also pulling on the Moon. If so, he wondered, why didn't the Moon fall to the ground too, rather than endlessly looping around Earth in its orbit?

It's sometimes said that Newton discovered gravity when an apple fell on his head, but this isn't true. In reality, it took him years to do all the maths needed to figure out how gravity works. And the apple didn't hit him!

"I am only a child playing on the beach, while vast oceans of truth lie undiscovered before me."

Isaac Newton (1643–1727)

Clockwork model of the solar system

Newton believed the Universe worked like clockwork, with the movement of the planets governed by simple mathematical laws.

The force of gravity

Galileo had found that a cannonball curves back to the ground as it falls because the force of its weight pulls it away from a straight line. Newton wondered what would happen if the cannonball went so fast that the curve of its path was even gentler than the curvature of Earth. The object would keep falling without ever landing – it would orbit Earth. With a stroke of genius, Newton saw that this is precisely what the Moon is doing, always falling but never landing. Pulled by the force of Earth's gravity, it is continually falling without getting any closer to Earth. It was just like Galileo's cannonball, but on a gigantic scale.

Across the Universe

Newton's next insight was to see that the Sun's gravity keeps the planets trapped in orbit around it for just the same reason. He realized all objects must exert a force of gravity on each other, the strength of which was in proportion to their combined mass. The Sun is so massive that it pulls whole planets towards it. Now Newton had enough clues to figure out why planets moved in the ellipses Kepler had discovered. The force of gravity must fade with distance, causing planets to slow down as they get further away but speed up as they got nearer (just like the ball and string on page 47).

Newton's GOOD points...

...and his BAD points...

Without doubt, Newton was a genius. His work on gravity established three "laws of motion" that describe how forces govern the way everything in the Universe moves, from atoms to planets. But Newton was also foul-tempered, nasty, and *very* eccentric...

- He was hugely intelligent and hard-working.
- He discovered how and why the planets move around the Sun, solving a centuries-old mystery.
- He founded the science of physics and figured out its most important laws.
- He invented a whole new branch of maths, now called calculus.
- He explained momentum and inertia.
- He discovered that white light is a mixture of different colours.
- He invented the reflecting telescope.
- He invented the milled-edge coin.
- He invented the catflap.

- He hated other people and worked alone.
- He made enemies easily and was spiteful.
- He wasted lots of time trying to find a recipe for gold (which was impossible).
- He used the Bible to calculate that God had created the world in the year 3500 BCE.
- He went through his great book the *Principia*, systematically removing every reference to the scientist Robert Hooke, whom he hated bitterly.
- He once told his mother and stepfather that he was going to burn down their house and kill them.

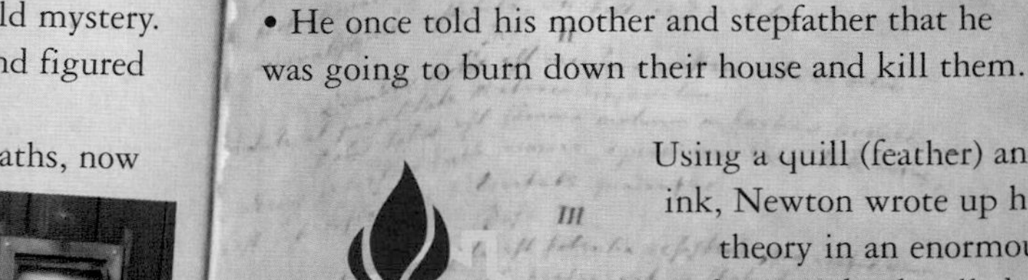

Using a quill (feather) and ink, Newton wrote up his theory in an enormous, handwritten book called the *Principia* (shown here). He made it as complicated as he possibly could – just to be difficult.

Where on Earth?

Up until the Middle Ages, the majority of people rarely travelled any great distance. Only armies and traders ventured more than a few miles or so from their home base. Maps were rare – instead, travellers used written instructions or remembered routes with landmarks such as rivers, mountains, or towns along the way.

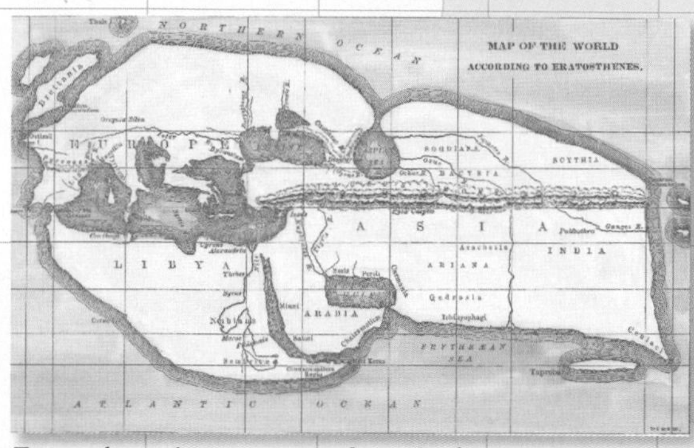

Eratosthenes's map centred on the known world but details of lands to the north and south were limited.

Anybody got a map?

Before you go anywhere, you need to know where you're starting from. Having worked out how big the world was, the Greek astronomer Eratosthenes tried to draw it as a flat map using a series of horizontal and vertical lines called lines of latitude and longitude. By making these into a grid, he could plot known landmarks and coastlines. Although the Greeks could work out latitudes quite well, longitude was more difficult and his map, though good for the Mediterranean area, was hopeless for the rest of the world.

The equator has a latitude of 0 degrees. It encircles Earth at an equal

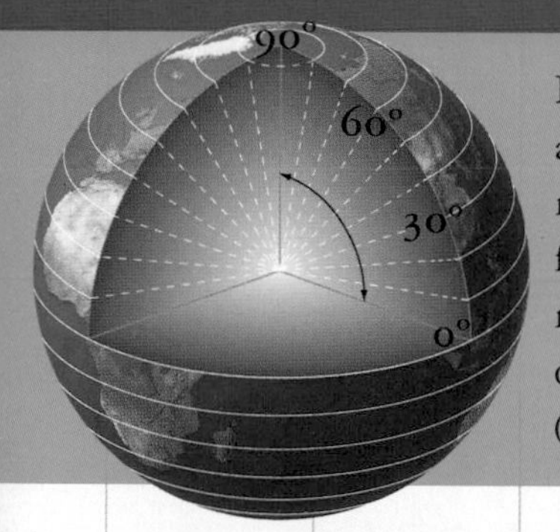

Lines of latitude tell you how far up or down you are on a map. They are horizontal lines that run parallel to the equator. Because latitudes are measured as angles upwards or downwards from the equator, they increase from 0 degrees to 90 degrees at each pole. On maps, latitudes are usually marked as being north or south, for example, 30 degrees N; or positive or negative depending on whether they are above or below the equator (–30 degrees is the same as 35 degrees S).

LATITUDE

Pole Star

Find your latitude...

As we saw earlier, you can find your latitude by measuring the height of the Pole Star. Hold your hands out at arm's length with your fingers level to the horizon. Each four-finger width is roughly equal to 15 degrees, so by counting how many fingers it takes to reach the Pole Star you can work out your latitude. You can use Sigma Octantis as the pole star in the Southern Hemisphere.

Astrolabe

Greek and Arab sailors and astronomers measured latitudes with a device called an astrolabe. This consists of a plate marked with a calendar, a map of the heavens, and degrees or hours around the rim. A movable plate marked with key stars and the Sun's path during the year fitted on top. By lining up the horizon and the Sun or a star using the pointers you could calculate your latitude or read off the date and time if you already knew where you were.

LONGITUDE

Lines of longitude measure how far left or right you are on a map. They run vertically from the North Pole to the South Pole. Because the equator is a circle, we can use this to divide Earth into 360 lines of longitude. They are measured from a prime meridian (0 degrees) that runs through London. Lines of longitude are measured as being either east (positive) or west (negative) of this line. The maximum number of degrees you can go east or west is 180.

0° 20° 40° 60° 80° 100°

The point where east meets west lies in the Pacific Ocean and is called the International Date Line. Go one way and it's yesterday, go the other and you find yourself in tomorrow. Who says you can't time travel?

… but it's harder to find your longitude

One of the first people to compile a list of places and their latitudes and longitudes was a Roman mathematician called Ptolemy. To work out longitude you need to be able to measure the time accurately – tricky when you don't have proper clocks. Ptolemy didn't do too badly, but the problem of longitude took 1700 years to solve.

It's time I found a better way to do this. Wish someone would invent a decent clock.

Ptolemy of Alexandria (90–168)

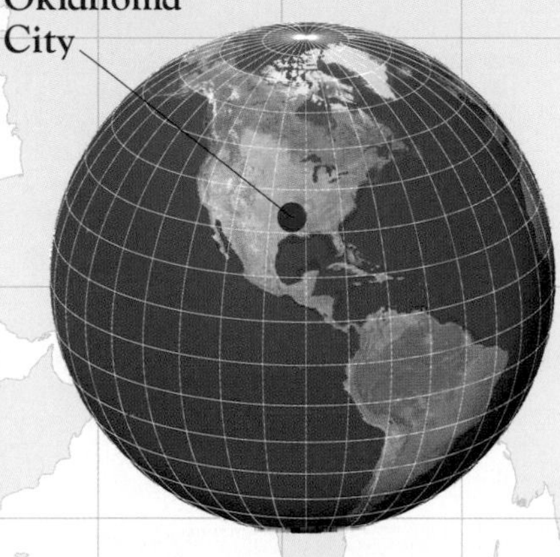

Coordinates

Mapmakers and navigators use latitudes and longitudes to mark points on Earth's surface. These pairs of numbers are called coordinates. This was a method that the Greek astronomer Hipparchus used to map the positions of stars and then adapted for land. First you look up or down from the equator to your latitude. Then you look east or west from the prime meridian until you reach your longitude. The point where the lines of latitude and longitude cross is your position. The coordinates of Oklahoma City would be written as 35°N, 97°W.

distance from the **North** and **South** poles.

Which way do we go?

Nobody knows where the idea of north, south, east, and west came from. Early man realized that the Sun came up in what we now call the east and set on the opposite side (west), having passed through its highest point halfway between the two (south). By working out where the Sun was at any time of day and heading towards or away from it, a traveller had a reasonably good idea in which direction he was going.

A compass rose

N

W E

S

Dry compass

Compass

Finding your way became much easier in the 11th century when the Chinese made the first compass. They found that if they stroked an iron needle with a magnetic rock and floated it on a bowl of water, it would align itself in a north–south direction. This worked well on a level surface, but couldn't be used at sea. A "dry" compass was invented in Europe around 1300. This was a needle that balanced on the tip of a pin, and was held level by a set of rotating supports called gimbals. The needle moved independently of the compass rose, which could be turned to show the direction you wanted to travel in.

All at sea

Finding your way out at sea was tricky. Most early sailors stayed within sight of the coast, where they could pick out harbours, river mouths, bays, and headlands. However, shore hugging could add weeks to a simple journey and might take you into hostile waters. At some point you had to head out to sea and find your way using only the Sun and stars to guide you.

It's too choppy to get an accurate reading on the astrolabe today.

To boldly go

Sailors setting sail would equip themselves with an astrolabe and tables of Sun and star positions to help them find their latitude. Once they'd hit the right latitude they would sail east or west to reach their destination, but it was all a bit hit or miss. Trying to measure angles on the rolling deck of a ship in high seas was difficult, and an error of just a few degrees could mean you missed land completely. Another problem was cloudy weather, when you couldn't see the Sun or stars. This was partly solved by the use of a compass to find your direction, but sailors still needed occasional clear skies to check their latitude.

We've hit the right latitude, Captain – do we go left or right from here?

This chart shows the coast of northwest Africa. The Portuguese were the first to go round the southernmost tip of Africa in their search for gold.

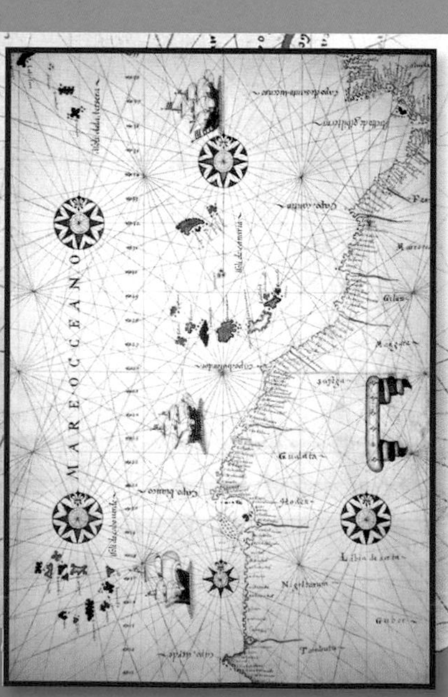

Portolan charts

Using a compass made a huge difference in the history of marine navigation. By the end of the 13th century, sailors had begun to draw the coastlines around Europe using compass bearings and latitude measurements. These "portolan" charts were criss-crossed with lines linking ports and landmarks all round the coast. If you set your compass to the right bearing and followed the line, you would eventually hit your destination.

Land ahoy, me hearties!

Tools of the trade

Are we nearly there yet?

Even if you were heading in the right direction, you still needed to work out your position using "dead reckoning". To do this sailors needed to know their speed and how much time had passed since they set off. Time could be measured using a sundial, astrolabe, or, less accurately, with a sandglass. Speed was measured using a chip log (see right) or by throwing barrels off the bow and timing how long it took them to pass the stern. Multiplying speed by time gives the distance travelled. Details of the course taken and how far the ship had travelled each day were entered into a "log" book. However, strong currents and winds could give you misleading results.

I reckon that sundial is slow – we should be in Genoa by now.

Follow that bird!

Crows were sometimes used by sailors as a navigational tool. In bad weather, they would be released from their cage up on the mast (the crow's nest) and the sea-hating birds would head straight for the nearest land.

Chip log

The chip log was a quarter circle of wood attached to a line that had been knotted at intervals of about 14.4 m (47 ft). It was thrown over the stern and a sailor counted how many knots went over the side in a given time (28 seconds). From this, a sailor could estimate the ship's speed, which is still given in "knots", or nautical miles per hour.

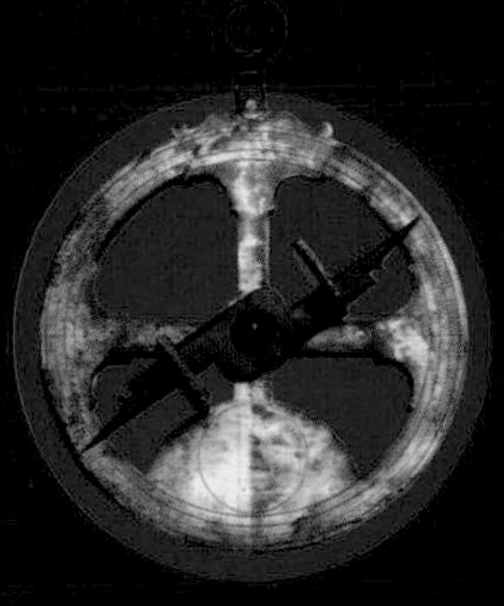

Marine astrolabe

The marine astrolabe was a heavier version of the astrolabe with holes cut out to stop it blowing in the wind. It had a rotating pointer to measure the altitude of the Sun or a star. The sailor would move the pointer until the Sun's light shone through holes at both ends, or the star could be seen. Then he would read off the angle of the pointer from a scale along the edge of the astrolabe. The angle could then be looked up in astronomical tables.

The quadrant

The quadrant was a quarter circle of wood or brass, marked along its curved edge with a 90° scale. A plumb bob hung from the centre of the instrument. The top edge was lined up with a star, and its angle above the horizon was measured from where the plumb bob crossed the scale.

The cross-staff

The cross-staff was a length of wood marked with a degrees scale. Another shorter piece of wood called a transom fitted across the staff at right angles, and could slide up and down. With the staff resting on his cheekbone, the navigator would move the transom until one end was on the horizon and the other was on the Sun or a star. He could then read his latitude on the degrees scale.

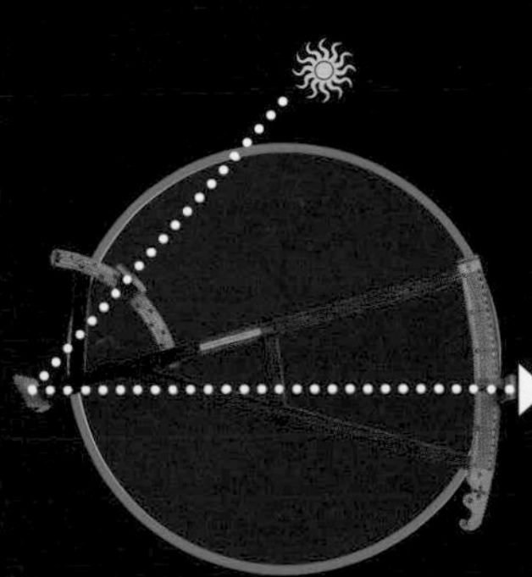

The backstaff

The backstaff was a similar instrument, but the sailor didn't need to look at the Sun, which could damage his eyes. He moved a sliding piece of wood on the small arc until it cast a shadow across a slit in the horizon vane. He then lined up a moveable slit on the large arc to see the horizon. He could then calculate the latitude using scales on the small and large arcs.

LONGITUDE

For centuries the biggest problem in navigating at sea was finding your longitude. To do this you need to know two things: the time where you are and the time back home.

EARTH ROTATES on its axis through 360 degrees every 24 hours, which means that for every 15 degrees you travel east, your local time moves one hour ahead (or one hour behind if you're travelling west). You can use this time difference to work out your longitude. For example, if you know it's 12 noon in London but your watch says it's 7 a.m., you must be five hours west of London. Multiply 5 by 15 and you get a longitude of 75 degrees – so you are somewhere on a line that runs through New York.

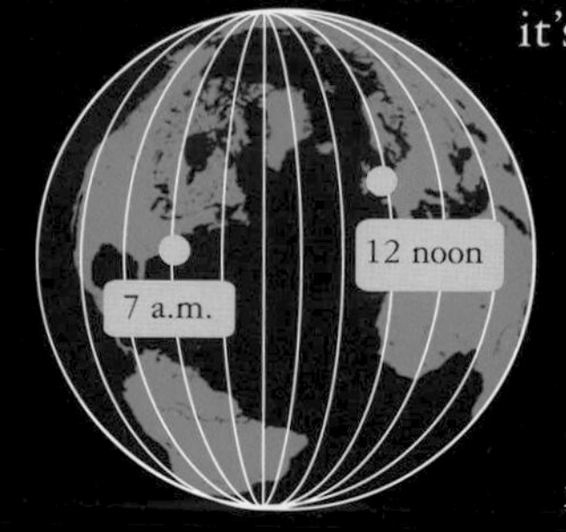

Where are we???

No idea!

Can you see the Moon?

It's time we got this problem sorted out!

Until the mid-SEVENTEENTH CENTURY there was no way to measure time accurately at sea, so calculating how far east or west you'd sailed was difficult. Several explorers and mathematicians put forward ways to measure time using observations of the **Moon** and **planets. These methods had their limitations** – they were impossible to use by day or on cloudy nights, and predicting the Moon's orbit required lengthy calculations. **In 1530, the Dutch mapmaker GEMMA FRISIUS** suggested using a clock to find longitude. It would be set on departure and then compared with a reading of local time from an astrolabe. Although this principle was correct, the pendulum clocks of the period always lost time over long voyages in rough seas.

I think this might do the trick – now give me the money!

Harrison solves the clock problem

In 1714 the British Government offered a prize of £20,000 to the first person to devise a reliable method of determining longitude at sea. The challenge was taken up by John Harrison, a carpenter and clockmaker. In 1735 he produced a clock, called H1. It used a pair of rocking bars instead of a pendulum to keep time. Although it kept time accurately during a trial, Harrison wasn't satisfied, and produced two more clocks, H2 and H3, before coming up with a design like a pocket watch. Watches use an oscillating balance wheel to keep time. He added jewels as bearings, and submitted it as H4 in 1759. On an 81-day sea trial to Jamaica in 1761, H4 lost only 5 seconds, well within the conditions for winning the prize. Despite this success, Harrison didn't receive his prize money until 1773.

H4

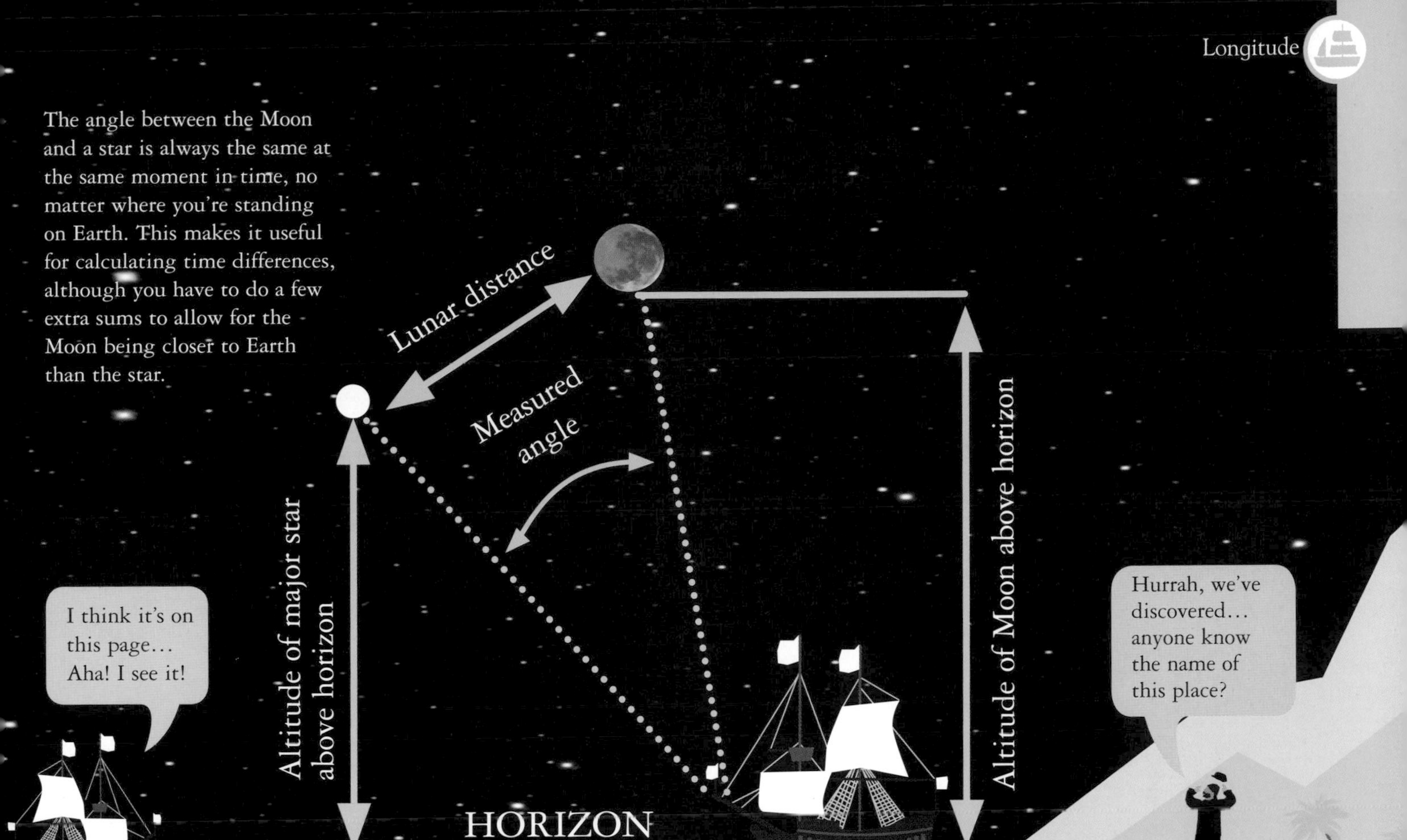

Lunar distance method

It took more than 100 years before accurate clocks became affordable to sailors. In the meantime, most used a new book of tables of distances between the Moon and nine key stars and the corresponding times these distances occured in Greenwich, London. The navigator would measure the angle between the Moon and a bright star close to it and work out the lunar distance. He then consulted the tables to find London time. After working out his local time using the altitude of the star, he could compare it with London time to calculate his longitude.

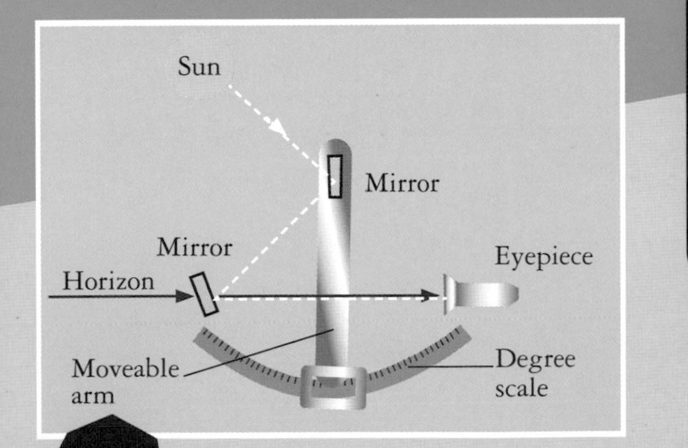

Captain James Cook was a great navigator. He was the first person to sail around the world in both directions, discovering Australia, New Zealand, Antarctica, and many Pacific islands in the process. He used dead reckoning, a sextant, and a compass to plot the course of his first voyage. On his second and third voyages he took a copy of Harrison's H4 clock, which allowed him to calculate his longitude more accurately and make detailed maps of his journey and discoveries.

James Cook, 1728–1779

The sextant was developed in 1757 by John Campbell. The sextant was similar to the quadrant but used mirrors to line up the horizon with light from a heavenly body. The two were aligned using a moveable arm that slid along an arc marked in degrees, which told you the angle. The sextant could be used to calculate both latitude and longitude accurately, and had the advantage that you didn't have to look at the Sun directly.

Mapping the world

I THINK I MAY HAVE MADE A MISTAKE...

Claudius Ptolemy (Ptolemy of Alexandria, 90–168 CE)

Making maps relies on mathematics. The hardest part is to turn a spherical world into a flat map. Because of this, all maps have problems, but there are many ways to get the view you need.

Off the map

The Roman mathematician Ptolemy compiled a list of latitudes and longitudes and turned them into a map. However, his map was wrong – he thought the world was much smaller, which made all his coordinates incorrect. Many sailors knew the map was wrong, but even when they corrected the figures they were reluctant to sail west because they knew their supplies would run out before they hit land again.

WHERE DID THAT CONTINENT COME FROM? IT'S NOT ON THE MAP!

Christopher Columbus

Despite knowing that Ptolemy's figures were probably wrong, in 1492, Christopher Columbus sailed west to find a quicker route to the riches of the East Indies. However, if he'd known that the distance from the Canary Islands to Japan was 19,600 km (12,200 miles) rather than the 3700 km (2300 miles) he'd calculated, he might never have left port. And he had no idea that America lay between him and the East Indies.

Archimedes 287–212 BCE

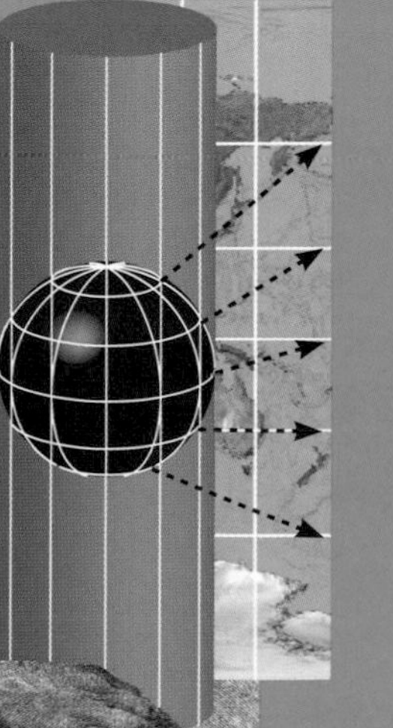

From round to flat

How do you draw a flat map of a round Earth? Over 2000 years ago, the Greek mathematician Archimedes discovered that the surface area of a sphere is equal to that of an equally tall cylinder enclosing it. Later mathematicians figured out various ways of "projecting" the points on a sphere onto taller cylinders around the sphere. Unroll the cylinder and you get a rectangular map.

Gerardus Mercator, a Flemish map and globe maker, used the cylinder technique to make a map of the world in 1596. There are a number of problems with this type of "projection", which stretches the continents in an east–west and a north–south direction. The distance between lines of latitude increases the further north or south you go, and it becomes difficult to measure distances near the poles – in fact you can't map the poles using this method. Also, it doesn't give a true idea of land areas: Greenland and Antarctica both look much bigger than they really are and they're stretched into odd shapes.

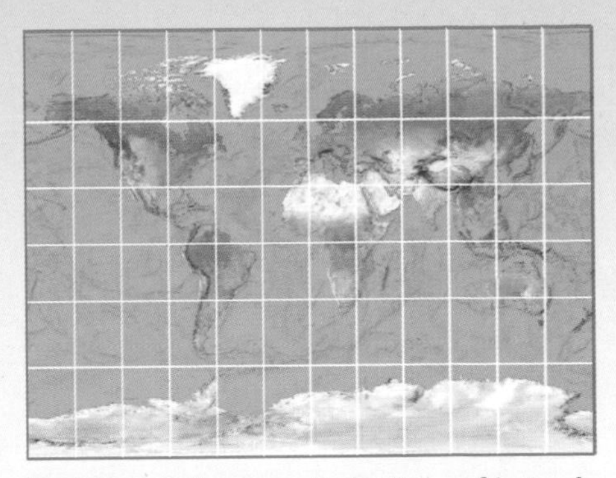

On Mercator's map, all the lines of latitude and longitude appear as straight lines.

Mercator's map proved very useful to marine navigators as it meant they could plot their course in straight lines. Until they could measure longitude accurately they used to sail according to compass directions. Although straight lines may look the shortest distance on a flat map, the shortest journeys between two distant points on a globe lie on what are called **great circles.**

GREAT CIRCLES

Great circles split Earth into two equal halves. The shortest route between any two places is always along the great circle connecting them. On a Mercator map, the shortest distance between London and New York looks like a straight line, but planes usually travel on a great circle that goes northwest over Ireland, passes south of Greenland, and then goes down through Canada.

Different projections

Because Mercator's map distorts the shape and size of many countries, map makers have come up with many different projections to suit the needs of the person using the map.

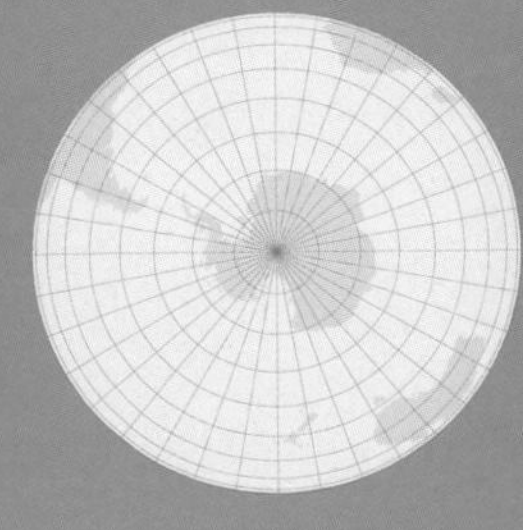

Azimuthal maps are circular, taken from one point on the surface, often one of the poles. They have also been used for maps of the stars since the time of Hipparchus.

Interrupted maps cut the world into sections called lobes. These preserve shapes and areas but are difficult to look at if they are cut into too many pieces.

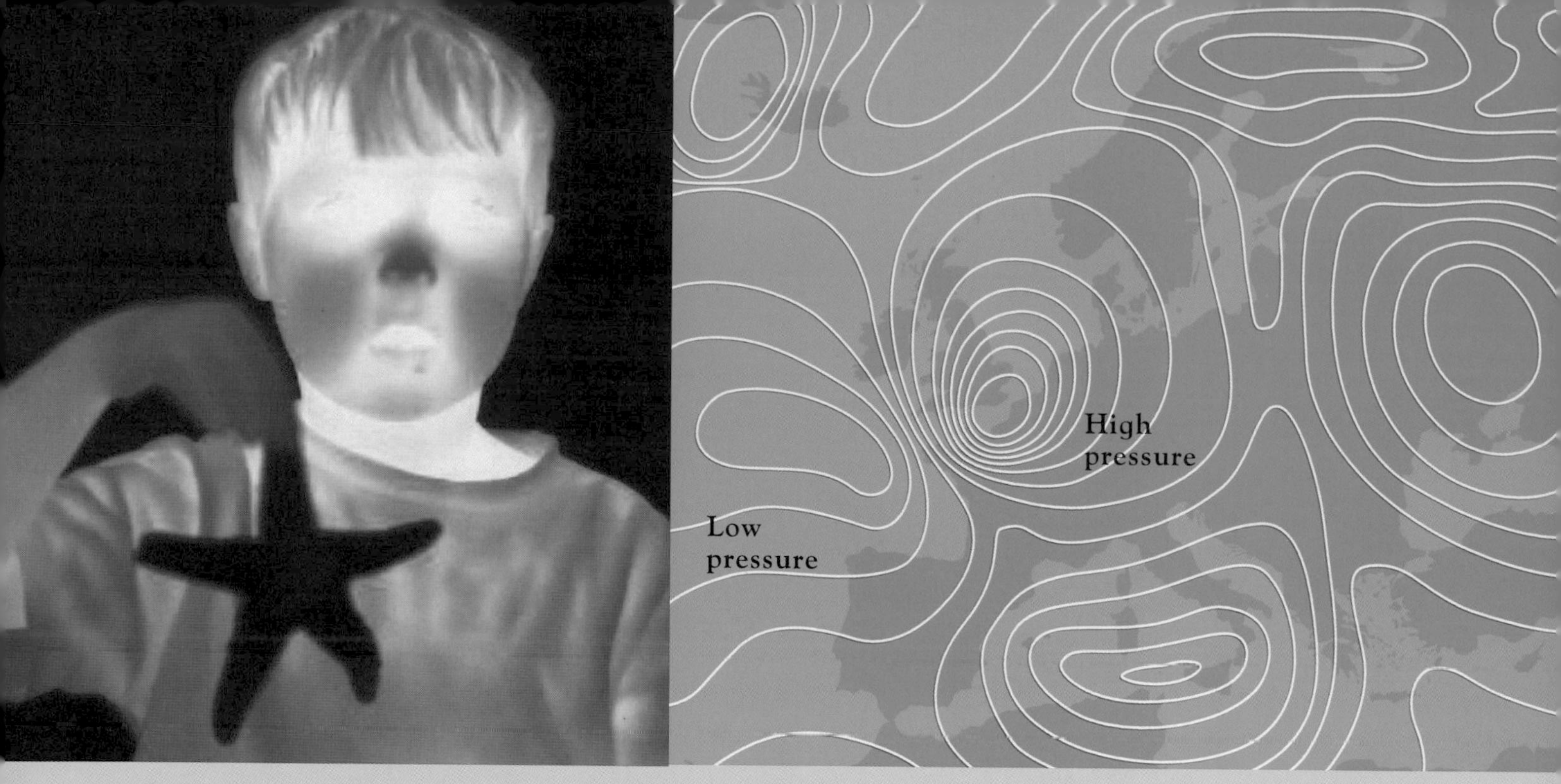

MODERN measuring

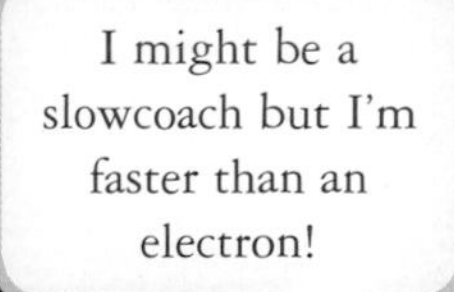

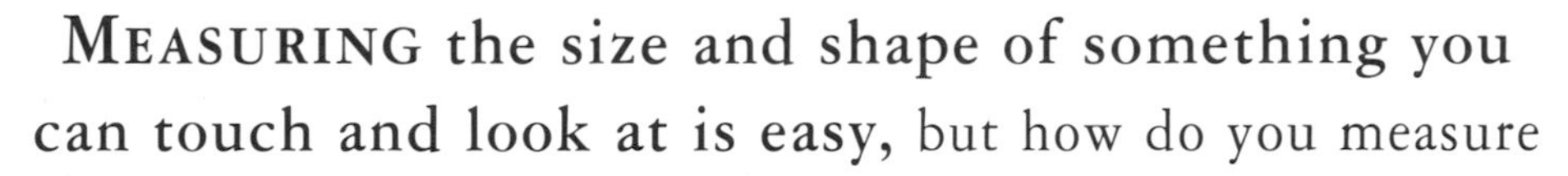

MEASURING the size and shape of something you can touch and look at is easy, but how do you measure things you can't even see, like **HEAT or SOUND**? And how do you **measure** things that are so **BIG** ***or so small that you can barely imagine them*** – such as an **ATOM** or a **GALAXY**?

The brilliant discoveries made by **GALILEO** and **NEWTON**, who we met in the last chapter, marked the beginning of the **age of science. They were the first true scientists,** not content merely to dream up theories but determined to check their ideas with meticulous **EXPERIMENTS** and **MEASUREMENTS.**

More scientists followed. They built powerful **microscopes** and **telescopes** to peer further into the unknown. **They invented devices to detect and measure heat, light, pressure, and sound.** And they discovered **ELECTRICITY, ATOMS,** and amazing new forms of energy that had been around us all along but were invisible. ***It was a revolution and it changed the world forever. None of it would have been possible without*** MATHS.

This is **science**, and **science is MEASUREMENT.**

Hot *and* cold

Ice is cold and fire is hot. We can tell this just by standing near them. But how do you measure how **hot** or **cold** something is? ***The answer is to take its temperature.***

100 NONILLION K
Temperature at Big Bang

5.5 TRILLION K
Hottest temperature reached in laboratory

22,000 K
Surface temperature of the star Bellatrix in Orion

6000 K
Temperature at centre of Earth

5800 K
Temperature of Sun's surface

380 K
Surface temperature of Moon facing Sun

373 K
Water boils

330 K
Hottest temperature reached on Earth

310 K
Average body temperature

What is heat?

Heat is a form of ***energy***. It comes from the ***movement of atoms and molecules***. The faster they move, the hotter they are. The temperature of an object tells us how fast its atoms or molecules are moving. Cold is a lack of energy – the atoms and molecules are not moving very fast. As they slow down, they become colder and colder. Eventually they stop altogether. We call this temperature absolute zero.

THIS STARFISH FEELS A BIT CHILLY TO ME.

Heat pictures

Everything has a temperature, but it is difficult for the human eye to tell whether something is hot or cold just by looking at it. Thermal cameras can pick up ***invisible infrared radiation***, which are the rays of heat energy given off by hot things, and turn it into a picture we can see. Because the amount of infrared an object gives out increases with temperature, it is easy to see warm things against cool backgrounds.

In this picture, the hottest areas are ***white*** and the coolest are ***purple***. Doctors use this type of picture to look for tumours, which get hotter than other parts of the body. Firefighters also use infrared cameras to look for people in smoke-filled buildings.

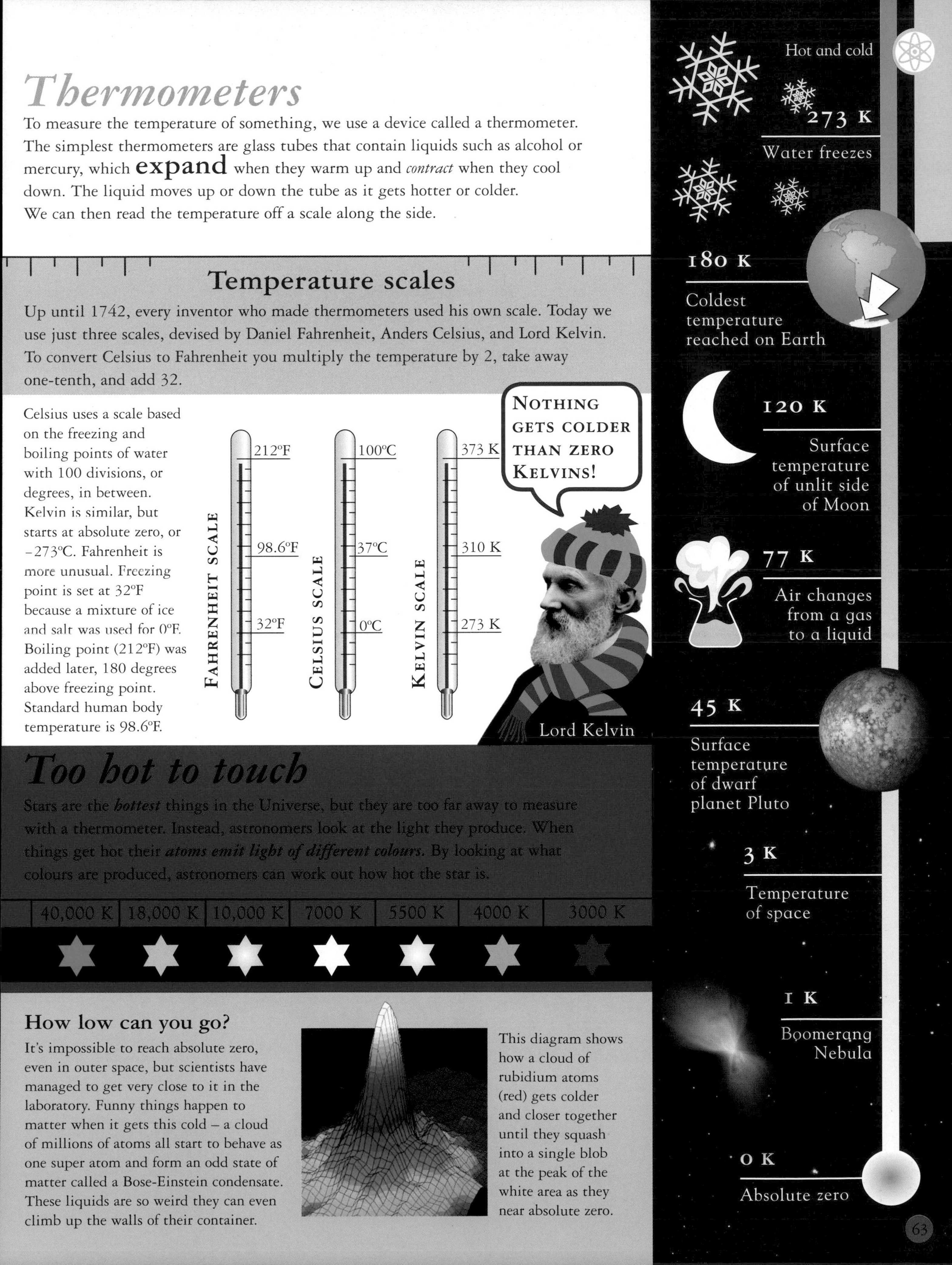

Thermometers

To measure the temperature of something, we use a device called a thermometer. The simplest thermometers are glass tubes that contain liquids such as alcohol or mercury, which **expand** when they warm up and *contract* when they cool down. The liquid moves up or down the tube as it gets hotter or colder. We can then read the temperature off a scale along the side.

Temperature scales

Up until 1742, every inventor who made thermometers used his own scale. Today we use just three scales, devised by Daniel Fahrenheit, Anders Celsius, and Lord Kelvin. To convert Celsius to Fahrenheit you multiply the temperature by 2, take away one-tenth, and add 32.

Celsius uses a scale based on the freezing and boiling points of water with 100 divisions, or degrees, in between. Kelvin is similar, but starts at absolute zero, or −273°C. Fahrenheit is more unusual. Freezing point is set at 32°F because a mixture of ice and salt was used for 0°F. Boiling point (212°F) was added later, 180 degrees above freezing point. Standard human body temperature is 98.6°F.

Lord Kelvin

Too hot to touch

Stars are the ***hottest*** things in the Universe, but they are too far away to measure with a thermometer. Instead, astronomers look at the light they produce. When things get hot their ***atoms emit light of different colours***. By looking at what colours are produced, astronomers can work out how hot the star is.

How low can you go?

It's impossible to reach absolute zero, even in outer space, but scientists have managed to get very close to it in the laboratory. Funny things happen to matter when it gets this cold – a cloud of millions of atoms all start to behave as one super atom and form an odd state of matter called a Bose-Einstein condensate. These liquids are so weird they can even climb up the walls of their container.

This diagram shows how a cloud of rubidium atoms (red) gets colder and closer together until they squash into a single blob at the peak of the white area as they near absolute zero.

Measuring *energy*

Energy is what makes everything around us happen. When we use energy it doesn't disappear – it simply changes from one form to another. We can't always see these changes, but heat, light, sound, and movement are all proof that energy is being used.

Newtons

Newtons, named after Isaac Newton, are actually units of force, but you need them to explain joules, the units of energy. A force is simply a push or a pull. The pull of gravity on a small apple when you drop it is about 1 newton (1 N).

OUCH!!
Newtons hurt!

Joules

A joule (J) is the amount of energy it takes to lift something with a force of 1 N (like an apple) one metre. It is also the amount of energy released if the same apple falls 1 m to the ground. Even when you're sitting still you emit 100 J of energy as heat every second.

JAMES PRESCOTT JOULE

OOOPS!

1 The Sun emits 386 trillion trillion watts.

2 The total amount of solar power hitting the atmosphere has been measured as 174 quadrillion W.

3 Of this, 89 quadrillion W is absorbed by land and sea.

4 The rest is reflected back into space.

An enormous quantity of solar energy hits Earth every day. Satellite technology tells us that only about half of it actually reaches the surface. The rest is reflected straight back into space. But we still receive a massive amount of free energy. The energy that reaches Earth every hour is about equal to the total amount of energy that humans use in a year.

How much POWER?

1 W
Hummingbird flying

30 W
CD player

80 W
Person reading

300 W
Food blender

380 W
Person walking

Energy and power

Power is the rate at which energy is produced or used. A moped and a monster truck may have the same amount of energy stored in their fuel tanks, but the truck can use its energy faster than the moped, so it is more powerful. Power is measured in watts and horsepower.

Energy is what makes things change, such as when ice turns to water.

Watts

The watt, named after Scottish inventor James Watt, is a unit of power rather than energy. One watt (W) equals one joule of energy per second. A 100 W light bulb, for instance, uses 100 J of energy every second, just like a person.

WATT A POWERFUL IDEA!

Horsepower

The horsepower is an old unit that James Watt used to measure the power of steam engines before the watt was devised. Watt based it on the pulling power of a horse, but he got his sums wrong – horses usually pull less than one horsepower.

ONE HORSEPOWER =735 W

(or 746 W in non-metric horses)

Body energy

Every day we power ourselves up by eating food. Food is full of energy, which we measure in kilocalories (kcal). One apple contains 55 kcal, which is equal to 230,000 J. This amount of energy would keep a 100 W light bulb going for about half an hour. An active, healthy 10-year-old uses around 2000 kcal, or 8.3 million joules of energy a day.

Using energy

We all use energy, but how efficiently do we use it? It might be easier to go to school in a car, but it uses more energy than if you walk or cycle. This is how many kcals it would take to travel 1 km (0.6 miles) to school:

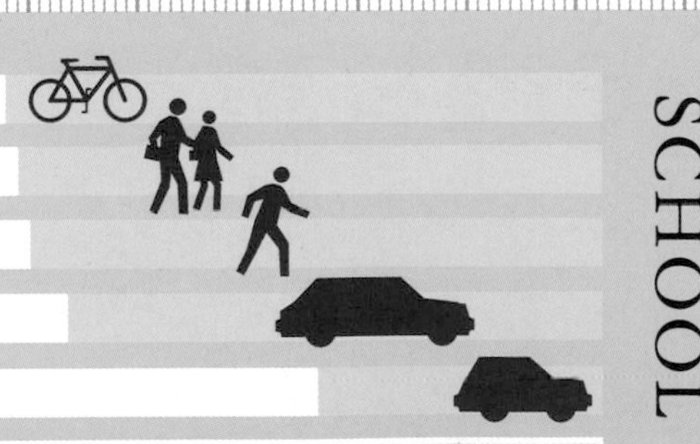

745 W
Person running

3000 W
Lawnmower

468,000 W
Racing car

4.5 million W
Train engine

11 billion W
Shuttle takeoff

70 quintillion W
2004 Indonesian earthquake

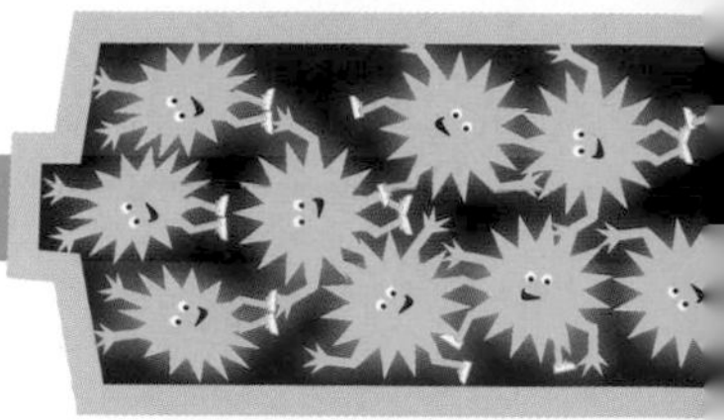

Electricity

It's invisible energy at the flick of a switch, but it's deadly too. We have to know how to measure and control electricity to put it to good use. Getting that wrong can have SHOCKING results!

Where does electricity come from?

Electricity comes from electrons, tiny particles that race around atoms carrying small packets of electrical charge. Normally, they stay put inside their atoms, locked in place by immense force. But sometimes they jump to another atom, taking their charge with them. When electrons gather together in one place, they produce what we call static electricity or simply "static" (measured in coulombs). When they move about, they make current electricity (measured in amps), which we usually refer to as "current".

Electron stampede!

When static builds up, it has a lot of potential to do something – like giving you an electric shock. We call this potential "voltage" (measured in volts). Connect something with lots of electrical potential (a thundercloud, say) to something with none (such as the ground) and it's like opening a dam. The static charge races from atom to atom in a sudden flood of current. A bolt of lightning is like a dam-burst of electrons, emptying from cloud to Earth.

Measuring current

Electrons are so amazingly small that they actually have zero size. That means that ***a lot*** of electrons can fit into a wire. In fact, it takes 6.2 quintillion electrons to make a measly 1 coulomb of charge. That's this many:

6,200,000,000,000,000,000 electrons

One amp of current (which is also a measly amount) is what you get when all those 6.2 quintillion electrons pass one point in a wire in a second.

Measuring voltage

Electrons are lazy and need a shove to get them moving. It takes energy to push them along – energy you have to supply with something like a battery. The more volts a battery has, the more it pushes the electrons, the more current they produce, and the more energy they deliver. It's easy to measure how much energy a bunch of electrons can carry through a circuit each second: it's just the voltage multiplied by the current, written in units called watts (W). Put more simply: watts = volts × amps

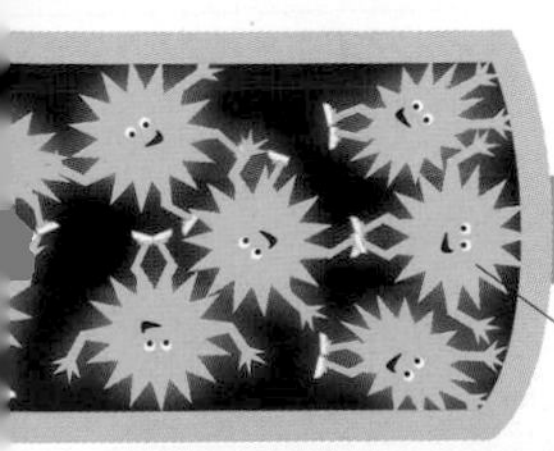

Zap!

A bolt of lightning is the biggest and most spectacular release of electric charge you'll ever see. A bolt of lightning carries as much as 100,000 amps of current (like 10,000 electric toasters all heating up at once). The smallest lightning bolts are as thin as pencil leads; the largest are as thick as a man's arm. Lightning heats the air to 28,000°C (50,400°F), making it expand so violently that it explodes, causing the bang that we hear as thunder.

How much power does it take to run these household products?

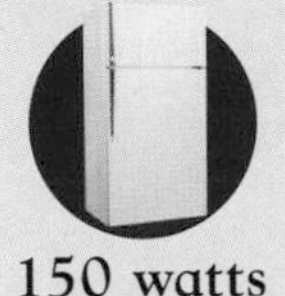

8 watts | 60 watts | 75 watts | 150 watts | 300 watts | 500 watts | 800 watts | 1500 watts | 2000 watts

The amount of energy an electrical device uses each second is its power, measured in watts. To find the total energy it uses, multiply its power rating by how long you use it. So, a computer with 300 watts of power running for 10 hours uses 3000 watt hours or 3 kilowatt hours (kWh) of energy. Electricity companies bill for how many kilowatt hours you use.

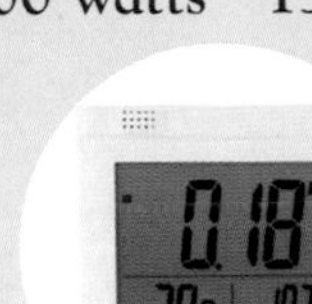
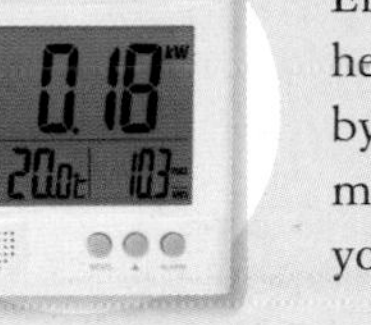

Energy monitors help save money by showing how much electricity you're using.

Shocking stuff

French priest Jean-Antoine Nollet (1700–1770) earned a place in history when he turned static electricity into current electricity for the King of France. First he stored up a stonking great electric charge in a glass-and-metal battery called a Leyden jar. Then he made 180 soldiers join hands and got the man at the end of the line to touch the jar. Zap! With a quick squirt of horizontal lightning, all the soldiers got a small electric shock and leapt in the air at the same time, much to the King's amusement. Nollet later repeated the experiment with 700 monks in a line 1 km (0.6 miles) long.

How fast does electricity go?

If an electric light flickers on in half a second, but your nearest power station is 100 km (60 miles) away, electrons must have shot through the cable at a breathtaking 720,000 km/h (450,000 mph) to reach your home, right? Wrong. You get instant power not because electrons race along wires but because they bump into each other, passing the charge all the way from the power station to your home. The electrons themselves drift along the wire about ten times slower than a snail.

I may be slow, but I'll get there faster than an electron.

A circuit is a closed loop through which electricity can flow.

Light fantastic

Light is a form of energy that travels in waves. Light is the reason we can see things. But there is more to light than meets the eye – there are other forms of light out there that we cannot see. We call all these types of light electromagnetic radiation. Light can be used to measure all sorts of things, using its wavelength, frequency, energy, and colour.

Wavelength

Radio waves

Light travels at a constant speed, but not all light waves have the same energy. Low energy waves have long wavelengths; high energy waves have short wavelengths.

Microwaves

Infrared

The number of waves that pass by in a second is called the frequency. The human eye can see only a very small range of wavelengths that we call visible light.

Isaac Newton discovered that **light** could be

Below red

The astronomer William Herschel was the first person to detect an invisible form of light. He was using a prism to split sunlight into a rainbow so that he could take the temperature of each colour. He had left the thermometer just beyond the red end of the spectrum and was surprised to see that the temperature was rising. The only explanation was that some invisible form of light lay beyond the visible spectrum. This became known as infrared radiation, which we feel as heat. Some animals, such as pit vipers, can see infrared, which they use for hunting prey.

Something's making me feel a bit hot under the collar...

Radar

Some wavelengths of light are used for measuring speeds and distances. Radar is a system that sends out radio or microwave signals and times how long it takes them to bounce back. From this we can tell how far away something is, or how fast it's going. Radar also compares changes in the frequency of the waves. The bigger the difference between the frequencies, the further the distance to the object. Aircraft use this method to check their altitude.

Radar can be used to map Earth's surface by sending radio waves from a satellite to measure the height of mountains.

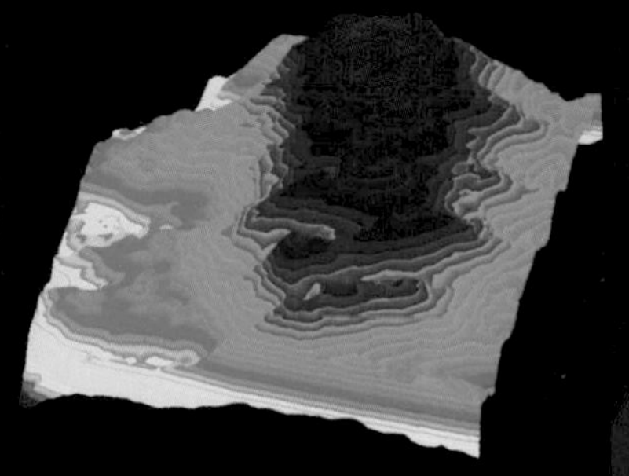

Spectroscopy

Spectroscopy is a technique that uses colour to measure things. When atoms are heated their electrons jump to higher energy levels. Eventually they fall back, but as they do, they emit light. Each element produces its own patterns of colour broken up by dark bands of

Rainbows

Every rainbow proves that light is made up of different colours. To see a rainbow you need to have your back to the Sun as you look into falling rain or a mist of droplets. As the sunlight enters a raindrop it is bent twice and splits into its individual colours. Red light is reflected to the eye at an angle of 42° to the path of the Sun's rays, and purple at 40°. We can measure these same angles if we look up from our shadow on the ground. If the light is bent three times, a secondary rainbow forms between 50° and 53°.

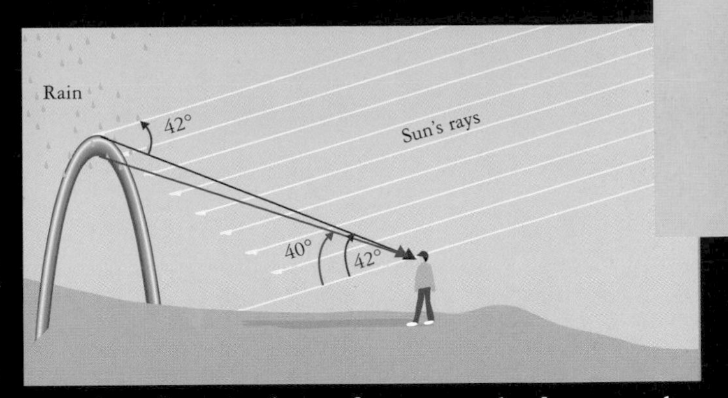

Visible light

Ultraviolet

X-rays

Gamma rays

split into different colours using a **glass prism.**

We know what the Sun is made from by looking at its spectrum and comparing it with the spectra of individual elements.

Solar spectrum

Potassium

Rubidium

Cesium

That UV's ruined my experiment, but it's given me a lovely suntan.

missing wavelengths, which act like a fingerprint. A spectroscope collects the light from an object and splits off each wavelength at a different angle so that the spectrum is spread out. By looking at the colours and the dark spaces between them scientists can work out what elements are in the object, how much of each element is present, and how hot the object is.

Beyond violet

Ultraviolet was discovered by the German physicist Johann Ritter, who noticed that silver chloride turned black when exposed to light. Invisible rays beyond the violet end of the spectrum were especially good at darkening the salts. Many insects can see ultraviolet, which helps them find nectar in flowers. Ultraviolet is also what makes your skin burn if you stay out in the sun too long.

Space

Light is the only way to measure objects in outer space. Telescopes pick up different wavelengths and frequencies of light, which we can use to measure the mass of stars and galaxies, detect which elements are present, determine temperatures, and find hidden things, such as black holes. The spiral structure of this galaxy has been revealed using telescopes that detect ultraviolet, infrared, and visible light. The purple dots are black holes and neutron stars.

Speed of LIGHT

Nothing travels faster than light, whether you're out in space or here on Earth. With a speed of just over a billion km/h (about 670 million mph) – that's about seven times around Earth in a ***second*** – light is pretty zippy. At that speed, light can get from New York to London faster than you can ***blink***. The weird thing is that no matter how fast you go to try

How do we know it goes this fast?

Several scientists, including Galileo, made attempts to measure the speed of light. The first person to get close to an accurate figure was Leon Foucault, who set up an experiment with a rotating mirror. He shone a light at the mirror, which bounced it to a stationary mirror and back again. The reflected light arrives a short distance away because the rotating mirror reflects it at different angle. If you know how fast the mirror is rotating and measure the distance between the outgoing and return light path, you can calculate the speed of light. Foucault got close, but not as close as Albert Michelson, who built a bigger and better version of Foucault's experiment. The most accurate speed we have is 299,792 km per second (186,282 miles per second).

Astronauts left mirrors on the Moon for scientists to fire lasers at, which has helped us get an accurate figure for the speed of light.

Foucault's Experiment

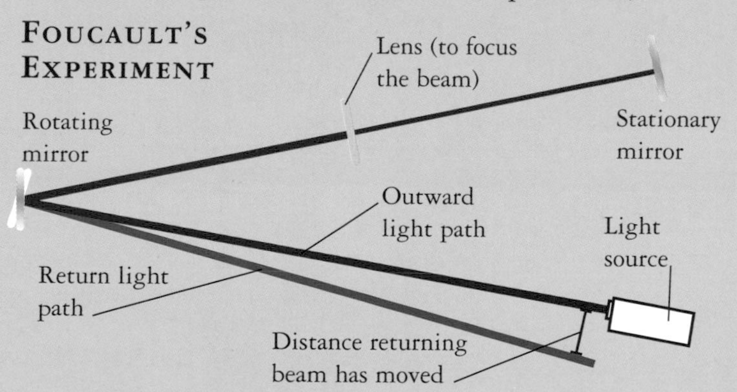

Light travels from New York to

Light years

The fact that light travels so far so fast means we can use it to measure the distance to the most remote stars and galaxies. If light travels at nearly 300,000 km/s, think how far it can go in a year. The answer is about 9.5 trillion km (5.9 trillion miles). We call this unit a light year. The closest star to our Solar System is Alpha Centauri, which is 4.3 light years away. That makes it about 41 trillion km, (25 trillion miles) away, which is hardly anything when you realize it is 30,000 light years to the centre of our galaxy and the furthest objects we can see are at the edge of the Universe, 46.5 billion light years away!

CENTRE OF OUR GALAXY
30,000 light years

and keep up with light, you can ***never*** catch it, or even get close to it. Light is always moving at **one billion km/h faster** than you are. It isn't only visible light that moves in this seemingly impossible way – all the other forms of electromagnetic radiation, from gamma rays to radio waves, do exactly the same thing.

Slowing down light

Back on Earth, light does slow down if it has to pass through something. It goes less than half as fast through a diamond, but will even travel (in the form of high-energy gamma rays) through a thick piece of lead at 120,000 km/s (72,000 miles/s).

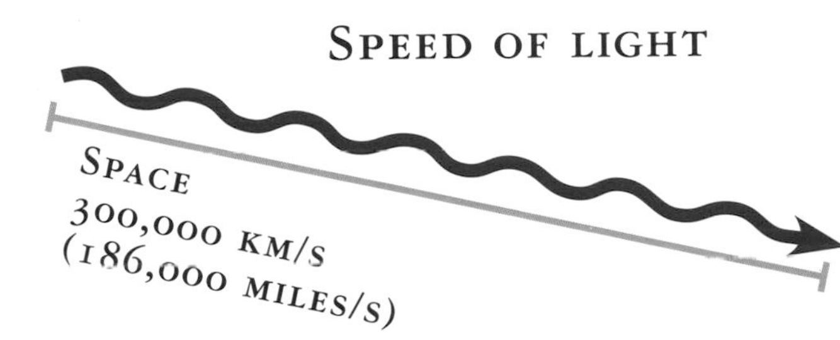

Glass
200,000 km/s
(124,000 miles/s)

Diamond
125,000 km/s
(77,500 miles/s)

London in *two hundredths* of a second.

Expanding Universe

We live in an expanding Universe. We know this because astronomers have discovered that many galaxies are moving away from us at great speed. As space stretches out, the wavelengths of light from galaxies are also stretched, producing what is called a redshift in their spectra. The effect this has is to move the dark lines in the spectrum of a galaxy closer towards the red end. By measuring how far the dark lines have moved, we can work out the age and distance of the galaxy. Galaxies with the biggest redshifts are found at the edge of the Universe. Objects with blueshifts, where the dark lines are shifted towards the blue end of the spectrum, are moving towards us.

Dark line in spectrum moves up towards red.

Redshifts not only tell us how fast the object's going, but also how far away it is.

Under pressure

We are all under pressure! As you read this book, the air around you is pushing on your body with a force equivalent to a 17-tonne weight. If your body were a hollow shell, you'd be crushed in an instant. BUT DON'T WORRY – we can't normally feel air pressure because our bodies usually push back with an equal and opposite force.

PRESSURE BELOW AND ABOVE WATER IN ATMOSPHERIC PRESSURES (BAR)

- 0.00001 BAR – Meteors 85,000 m (300,000 ft)
- 0.001 BAR – Weather balloon 50,000 m (165,000 ft)
- 0.1 BAR – Ozone layer 16,000 m (50,000 ft)
- 0.3 BAR – Aeroplane 8000 m (26,000 ft)
- 0.5 BAR – Mountain top 5000 m (16,000 ft)
- 1 BAR – Sea level
- 2 BAR – Diver –10 m (–33 ft)
- 30 BAR – Shark –300 m (–1000 ft)
- 300 BAR – Squid –3000 m (–10,000 ft)
- 500 BAR – Anglerfish –5000 m (–16,000 ft)
- 1000 BAR – Bottom of Pacific Ocean –11,000 m (–36,000 ft)

1 bar

WHAT IS ATMOSPHERIC PRESSURE?

Air is not empty space – it's packed with trillions of invisible gas molecules that move about all the time, bumping into each other and into other things. Trillions hit you every second, each giving a tiny shove. All the shoves add up to create pressure. The air molecules in Earth's atmosphere are pulled towards the ground by gravity, so air is densest near the ground and the pressure highest there. We measure air pressure in units called bars. The pressure at the bottom of the atmosphere (sea level) is 1 bar.

Forecasting the weather

One of the best ways to predict the weather is to measure air pressure. We use a device called a barometer to do this. Some barometers even come with a prediction printed on the pressure dial. When weather forecasters talk about "highs" and "lows", they mean areas of high and low pressure. High-pressure areas are usually calm and sunny, while areas of low pressure usually have bad weather.

In the heart of this barometer is a sealed can...

... as the air pressure rises and falls, the can shrinks or expands, making the pointer rotate.

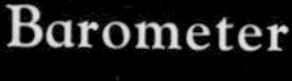
Barometer

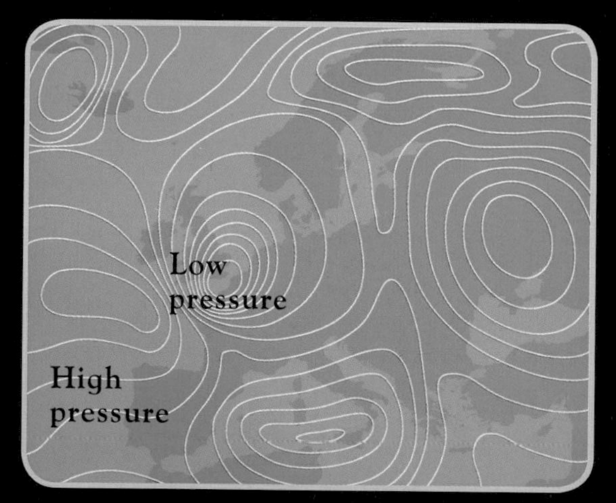

Weather map

The air molecules in areas of high pressure rush to fill up the areas of low pressure. This rushing air is what we call wind. And wind often picks up moisture, creating clouds and rain.

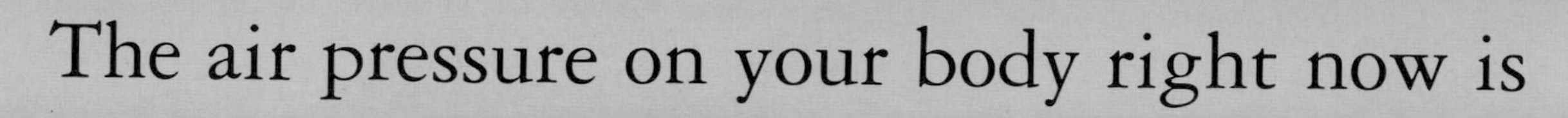
The air pressure on your body right now is

0.3 bar

Flying high

After a plane takes off, you can sometimes feel your ears popping. This happens because the air pressure in the cabin drops as the plane rises, but the air inside your ears stays at normal atmospheric pressure and pushes on your eardrums. The air pressure in the plane doesn't drop as much as it does outside it. If it did, the air would be too thin to breathe and everyone would suffocate. So the cabin is pressurized to keep the air breathable, but the pressure is a little lower than at ground level.

0.5 bar

Mountain top

Air pressure falls as you rise through Earth's atmosphere because the air molecules become less densely packed. At the top of a mountain, the air is so thin that it can make breathing difficult. You have to take much bigger mouthfuls of air with each breath to get the oxygen molecules your body needs.

Diving deep

Our bodies are perfectly adapted to atmospheric pressure on land, but what happens if you go scuba diving? Water molecules are much heavier and more densely packed than air molecules, which means they create much more pressure. You only have to dive 10 metres (33 feet) deep for the pressure on your body to double. To stop the sea from crushing your lungs, you have to breathe high-pressured air from a tank so that the pressure in your lungs matches the pressure of the water outside.

2 bar

I'm 10 m (33 ft) deep and the pressure is mounting...

The bends

Divers have to be very careful not to come out of the sea too quickly or they risk suffering a deadly medical condition: the bends. This is caused by the high-pressure air they breathe while underwater. Under high pressure, nitrogen molecules in the air supply start to dissolve in the diver's blood. If he rises too fast, the sudden drop in pressure causes the nitrogen to form deadly bubbles in his body, just as bubbles form when you open a fizzy drink.

Secret water squirter

Here's a practical joke to play on nosy parkers. It works thanks to air pressure.

1. Write "Do not open!" on a plastic bottle. Fill with water, screw on the lid, and make some holes in the bottom with a pin.

Take any pins out of the bottle. Don't worry, it won't leak – air pressure will stop the water coming out of the holes, as long as you leave the lid on.

2. Stand the bottle where a nosy parker might find it. The label will make them curious. They'll open the bottle and get a soaking!

Do not open!

When the lid is taken off, the air gets in and pushes from the top too, making the water squirt out.

equivalent to the weight of four elephants!

Can you hear me?

The world is full of sound. Sound is another form of pressure, caused by molecules bumping into each other and passing on their energy. These ripples of pressure travel in waves through the air like ripples on a pond until they reach our ears.

On the crest of a wave

The shape of the pressure wave tells you a lot about the sound – whether it's loud or soft, or high or low in pitch. The distance between the peaks of a wave is called the wavelength. The number of waves that pass by each second is called the sound's frequency.

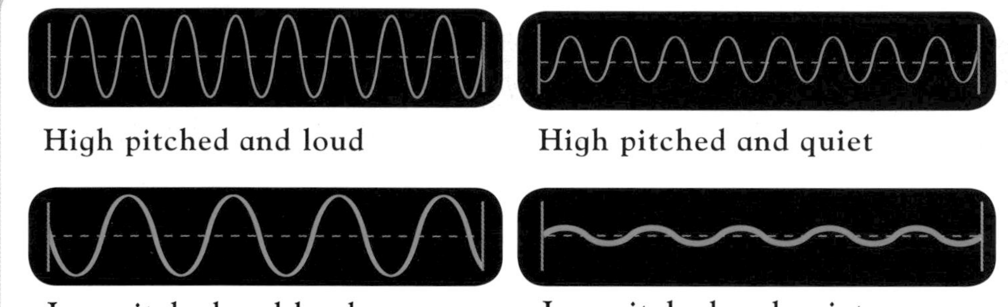

Frequencies

Frequencies are measured in units called hertz (Hz). The frequency tells us the pitch of the sound – waves with peaks that are very close together have a higher pitch than long, spread-out waves. Measuring the height of the wave tells us how loud the sound is. Tall waves sound louder than flatter waves.

0 dB Lowest level audible to humans
10 dB Leaves rustling
40 dB Person talking nearby
70 dB Main road
100 dB Road drill
110 dB Front row of rock concert
120 dB Hearing damaged
140 dB Jet engine 30 m away
155 dB Dragster engine
180 dB Krakatoa eruption 160 km (100 miles) away
188 dB Blue whale singing underwater
200 dB Space shuttle blasting off
210 dB 1 tonne TNT bomb
218 dB Pistol shrimp snapping its claw underwater
300 dB 1908 Tunguska meteor that hit Russia
300+ dB Asteroid strike that led to death of dinosaurs

0 50 100 150 200 250 300

Ding-dong decibels

We use decibels (dB) to describe the power of a sound pressure wave. A decibel is one-tenth of a unit called a bel, which was named after Alexander Graham Bell. Decibels are measured from the lowest sounds our ears can detect, such as a faint whisper. Every extra 10 decibels is a tenfold increase in sound pressure, so a 10 dB sound is ten times more powerful than a whisper, 20 dB is 100 times, 30 dB is 1000 times, and so on. We call a scale that increases in multiples of ten like this a *logarithmic scale*. Logarithms are useful for making really big numbers easier to work with. For example, the level at which our hearing becomes damaged (120 dB) is one trillion times, or ten multiplied by itself 12 times, greater than that of a whisper.

HELLO ALEX!

You don't need to shout – I'm only in the next room.

Alexander Graham Bell
Inventor of the telephone

Amplitude (how loud the sound is)

Wavelength

Not all sound waves are as smooth as this. Noise and speech are a jagged mixture of peaks made up of lots of different frequencies and pitches. Patterns of speech are sometimes so unique they can be used as voiceprints to identify people.

Pattern of human speech

Speed of sound

Sound waves are transmitted by molecules vibrating against each other, so sound will travel through liquids and solids as well as air. In fact, it usually travels faster through liquids and solids because their atoms and molecules are much closer together. In solids, sound travels faster through stiff materials than soft materials. Space is the quietest place in the Universe because sound cannot travel across a vacuum.

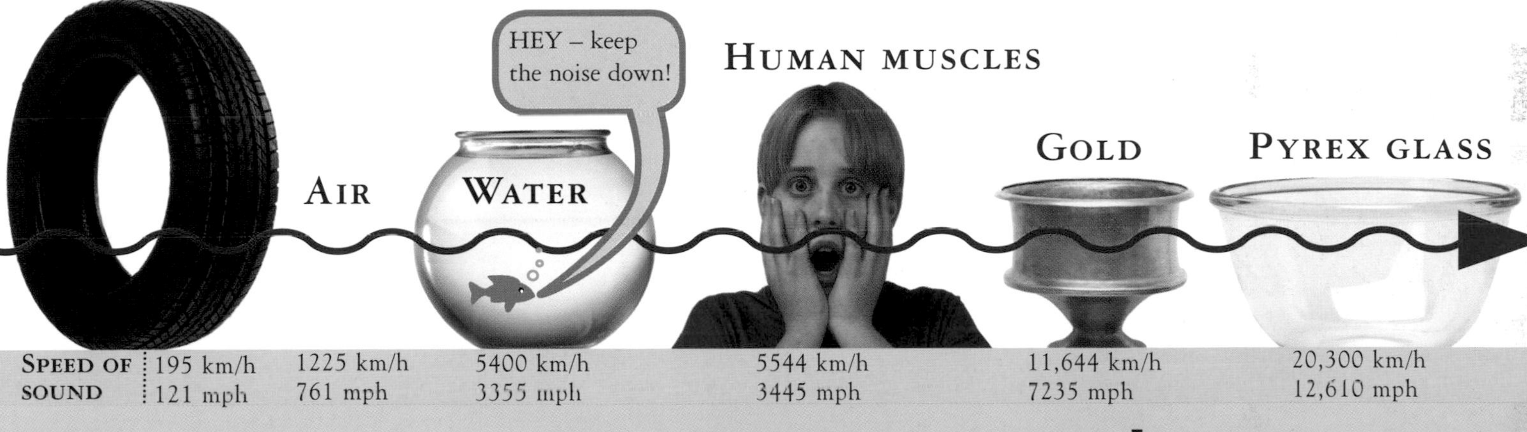

SPEED OF SOUND	Rubber	Air	Water	Human muscles	Gold	Pyrex glass
km/h	195 km/h	1225 km/h	5400 km/h	5544 km/h	11,644 km/h	20,300 km/h
mph	121 mph	761 mph	3355 mph	3445 mph	7235 mph	12,610 mph

Sonar

Because sound can travel through other materials it can be used to detect and measure things that we can't see. One of its chief uses is sonar. Sonar is used by boats and submarines to explore underwater and detect fish shoals. A sonar system works by sending out pulses of sound then capturing the echoes. If it takes six seconds for a signal to come back, then it has taken three seconds to reach its target and three seconds to return. Since sound travels at 5400 km/h (3355 mph) in water, we can then work out that the target is 4.5 km (2.8 miles) away. Whales, dolphins, and bats use the same system (echolocation) to navigate and find food.

Highs & lows

I can hit some really high notes.

Who says giraffes haven't got a voice?

The human ear can detect a wide range of frequencies from 25 Hz to 20,000 Hz, though it hears sounds between 1000 and 4000 Hz best. However, other animals can hear much higher or lower frequencies than we can. Bats, whales, and dolphins can transmit and hear very high frequencies, which they use for echolocation. Sounds with frequencies too high for human hearing are called ultrasound. Ultrasound is used by doctors to see inside the body. Noises with frequencies below our level of hearing are called infrasound. Elephants, giraffes, and hippos all use infrasound. Some elephant calls travel through the ground, and the vibrations are picked up through their feet.

The sound of music

People who are good at maths are often good at music too, but what has music got to do with maths? As the ancient Greeks discovered thousands of years ago, music is full of hidden ***mathematical patterns.***

A scale is a sequence of notes of rising frequency that progress in a pleasing way. The note at the top of the scale has exactly twice the frequency of the one at the bottom and sounds similar but higher. The interval between these two is called an octave.

1 2 3 4 5 6 7 8

An octave

C	D	E	F	G	A	B	C
262 Hz	294 Hz	330 Hz	349 Hz	392 Hz	440 Hz	494 Hz	524 Hz

Each musical note has a distinct frequency, shown here in Hertz (sound waves per second).

Measuring music

The Greek mathematician Pythagoras was one of the first people to find maths hidden in music. The story goes that Pythagoras was passing a blacksmith's when he became curious about the ringing notes made by hammers striking anvils. He decided to investigate. He discovered that an anvil twice as big made a lower musical note of exactly half the pitch. He'd found a mathematical connection between the size of the anvil and the pitch of the sound it made.

I've got it! It's all to do with the length of the string!

Pythagoras (575–500 BCE)

Strings and things

Pythagoras wondered if he could find a similar mathematical pattern in string instruments. Sure enough, it turned out that halving the length of a string resulted in a note exactly twice the pitch (one octave higher), because the shorter string vibrated twice as fast. Doubling the length produced a note of half the pitch (one octave lower). Pythagoras also found that by making a string longer or shorter in exact fractions, or by tightening it with carefully measured weights, he could create all the notes on the musical scale.

If a guitar's string is halved, the resulting note is one octave higher.

"Music is the pleasure the human mind experiences from counting without

Feel the beat

Put your hand to the left side of your chest and you'll feel your heart beating. It pounds around 60–70 beats per minute (bpm) when you're relaxed and up to 200 bpm when you're excited. All music has a rhythmic pulse just like a heartbeat. When you tap your feet or dance to music, you're synchonizing your body with the music's "tempo", or speed. Gentle music has a slow tempo of 60–70 bpm, like a relaxed heart, but energetic dance music may have a tempo of 200 bpm, like a heart pounding flat out.

The human heart produces a rhythmic pattern like a drumbeat.

Digital music

How can thousands of music tracks be stored on a tiny MP3 player? It's all thanks to numbers. When music is recorded, a computer system captures every sound wave – even for a full orchestra – by measuring the pitch (frequency) and loudness up to 100,000 times a second. These measurements are stored as strings of digits (numbers), which is why the music is described as digital. When you play the tracks, your computer or MP3 player turns the digits back into sound waves.

Sound waves can be measured using numbers. The wave peaks have the highest numbers and troughs the lowest.

Keeping time

When musicians play together in a band or an orchestra, it's essential they all stick to the same rhythm and tempo, like dancers keeping in step. There are several techniques to help them keep to the beat.

Conductor

Part of a conductor's job is to keep his orchestra of 50–100 musicians in time. He marks the beat with movements of his baton, allowing musicians to count beats when they aren't playing.

Drummer

In modern bands, no conductor marks time. Instead, the drummer helps the musicians measure out time by making the beat audible. The drummer is a bit like a conductor you can hear.

Metronome

Musicians playing alone sometimes use a device called a metronome to help them keep a constant tempo. A weighted rod ticks as it swings from side to side. Sliding the weight up or down alters the speed.

Conducting

Conductors don't simply wave their batons back and forth. They move them in specific patterns that show the rhythm of the music and tell the musicians when to emphasize a beat. The position marked with a one in the four diagrams above is where musicians make the beat strongest.

being aware that it is counting." Gottfried Leibniz (1646–1716), German mathematician

Modern TIMES

Time is a measurement we live in – an invisible ruler that rules our lives. Modern technology allows us to chop up and measure time in ever tinier fractions, but will technology ever allow us to warp time so we can zoom forward to the future or back into the past? Only time will tell.

KEEPING TIME

To keep time, all clocks and watches rely on what's known as a "harmonic oscillator" – a physical device that vibrates back and forth (oscillates) at a constant frequency.

Swinging weights (1650s–)

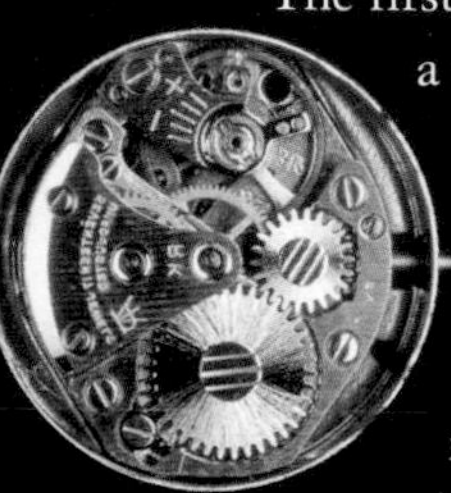

The first accurate clocks kept time with a swinging weight – a pendulum. Later clocks and early watches used a rocking bar or a rocking wheel to do the same job. This allowed the mechanism to be miniaturized, making the machine portable.

Quartz vibrations (1960s–)

Most modern watches keep time using tiny quartz crystals that shudder precisely 32,768 times a second. A microchip counts these vibrations and turns them into hours, minutes, and seconds.

Atomic vibrations (1990s–)

Atomic clocks use the vibration of electron particles inside atoms to keep time. They are accurate to one second in every 60 million years. Atomic watches receive daily radio signals from atomic clocks to make sure they always show the perfect time.

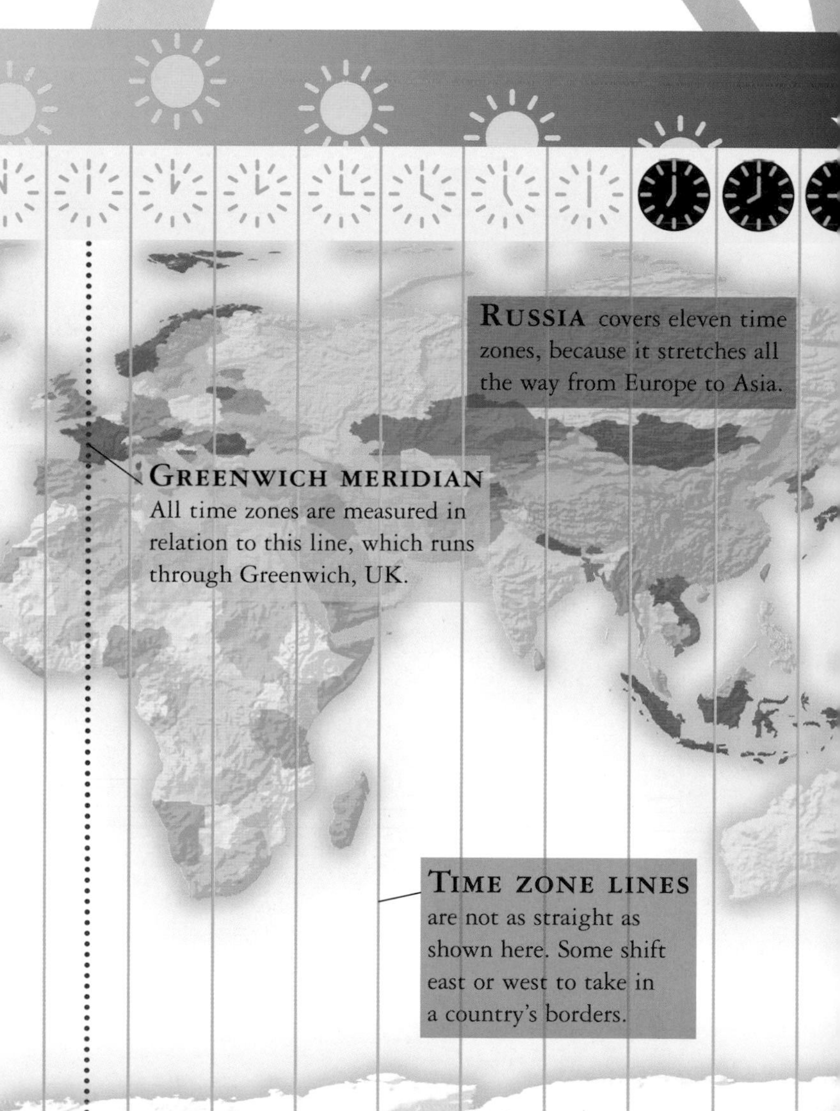

RUSSIA covers eleven time zones, because it stretches all the way from Europe to Asia.

GREENWICH MERIDIAN All time zones are measured in relation to this line, which runs through Greenwich, UK.

TIME ZONE LINES are not as straight as shown here. Some shift east or west to take in a country's borders.

TIME ZONES

Until the 18th century, most places on Earth measured time differently, setting their own local time using sundials. Now the entire world counts time the same way using Coordinated Universal Time (UTC). This divides the globe into 24 zones, each of which is an exact number of hours ahead of or behind London, England, where the time is called Greenwich Mean Time (GMT).

Walking the Planck

What's the smallest time you can measure? In prehistory it was the day. By the 16th century it was the second. Today, sat nav systems depend on clocks in satellites that are never more than a billionth of a second out (otherwise your car might end up on the wrong side of the road). But there's still room for improvement. The smallest time anyone will ever be able to measure is the "Planck time", which is 0.000000000000000000000 00000000000000000000005 seconds long and named after German physicist Max Planck. It's impossible to divide time into shorter measures than that.

0.00000000000000000000000000000000000000000005

International date line This imaginary line separates one day from the next. It curves around the islands of Kiribati so they can have a single time zone.

The Poles
The world's time zones meet at the North and South poles. By walking around the poles themselves, it's possible to travel through all the world's time zones in a matter of seconds.

Metric time

Why has time never gone metric? France tried it briefly after the 1789 Revolution. There were 10 days a week, 10 hours a day, 100 minutes an hour, and 100 seconds a minute. Months were named after seasons and weather, so your birthday might have been the 11th of Fog (October) or the 27th of Fruit (June). Metric time was hugely unpopular: everyone still got a day off a week, but a week was 3 days longer so that meant only one day off in 10!

Internet time

Instead of time zones, everyone in the world could use the same time but start and finish the day at different points. That's the idea behind Internet time. A day is made of 1000 units and time is simply written as a number from 000 to 999.

Time travel

How could we travel in time? American mathematician Frank Tipler (1947–) thinks we'd need to stretch space and time first using a giant spinning pipe. We could then chug around the pipe in a spacecraft, nipping forward to the future or back into history. The catch? Tipler's pipe might need to be 10 times heavier than the Sun, infinitely long, and running on negative energy!

Rod Taylor in the 1960s film *The Time Machine*.

Disaster!

Is a hurricane worse than an earthquake? Just how big an asteroid impact could the world withstand? Planet Earth has always been and always will be a dangerous place to live. We may marvel at the power unleashed by weapons of mass destruction, but they are puny compared to the violence of natural disasters.

The Torino scale

0	**No hazard:** Virtually zero chance of collision
1	**Normal:** A rock passing near with little cause for concern
2	**Meriting attention:** A rock whistling by but unlikely to hit
3	**Meriting attention:** A rock with a 1 per cent chance of hitting and causing limited localized damage
4	**Meriting attention:** A rock with a 1 per cent chance of hitting and causing regional devastation
5	**Threatening:** A rock, still some way off, that might cause serious regional devastation
6	**Threatening:** A rock some way off that might cause a global catastrophe
7	**Threatening:** A large rock nearby that poses a major risk of a global catastrophe
8	**Certain collision:** A rock that will definitely cause local damage or a tsunami at sea
9	**Certain collision:** A huge rock that will cause massive regional destruction or a tsunami
10	[illegible]

Astero-disaster

Dinosaurs probably became extinct when an asteroid (a gigantic space rock) smashed into Earth about 65 million years ago. But thousands of meteorites (ranging in size from car-sized boulders to tiny flecks of space dirt) happily strike Earth each year, often with no effect. Space scientists use the Torino scale to measure the danger that space rocks pose.

Shaking quakes

Earthquakes happen when the vast plates that make up Earth's crust suddenly rupture or jerk, shaking the ground. Because large earthquakes are massively more destructive than small ones, scientists use a special scale to measure them. Each step up the scale means the quake is 30 times more powerful than the last. So an earthquake that measures 8 isn't eight times more powerful than an earthquake measuring 1, it's 30 billion times more powerful! A scale of measurement that increases this way is called a "logarithmic" scale.

Moment magnitude scale

- 8+ **Great** Massive devastation, enormous loss of life
- 7 **Major** Huge devastation, major loss of life
- 6 **Strong** Widespread damage, fatalities possible
- 5 **Moderate** Damage likely, fatalities rare
- 4 **Small** Local damage possible
- 2–3 **Minor** Damage unlikely
- 1 **Not felt**

Kaboom!

Volcanic explosivity

An erupting volcano is the last thing you want to try measuring – unless you want to find yourself buried under a million tonnes of lava. So how do scientists safely compare volcanic eruptions? Standing at a safe distance, they estimate the volume of material spewed out from the top, how high it goes, and how long the eruption lasts. The bigger these numbers, the higher the volcano scores on the Volcanic Explosivity Index (VEI).

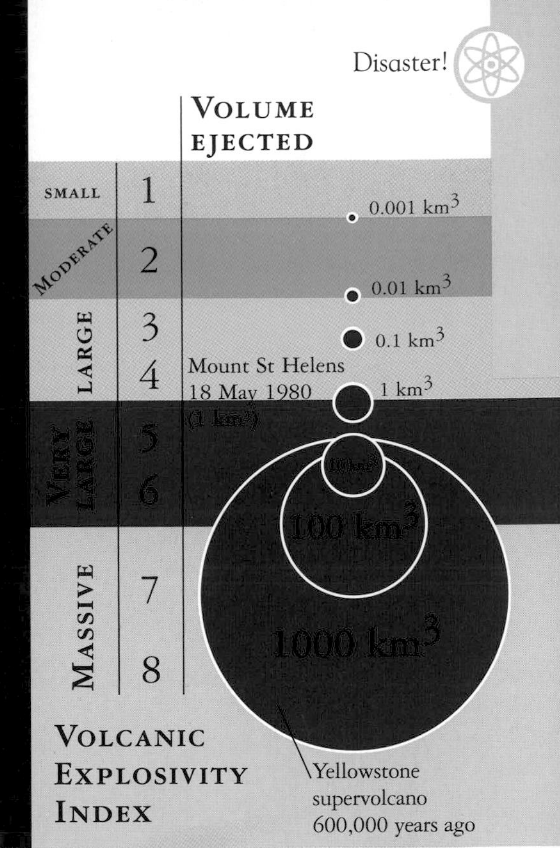

Hurricanes

Dolly, Katrina, and Andrew... hurricanes with friendly names sound harmless, but these spinning oceanic storms (also called typhoons) cause more devastation than any other natural disaster. A hurricane can release as much energy in two minutes as a nuclear bomb. Estimating the power of hurricanes is a vital part of assessing the danger they pose.

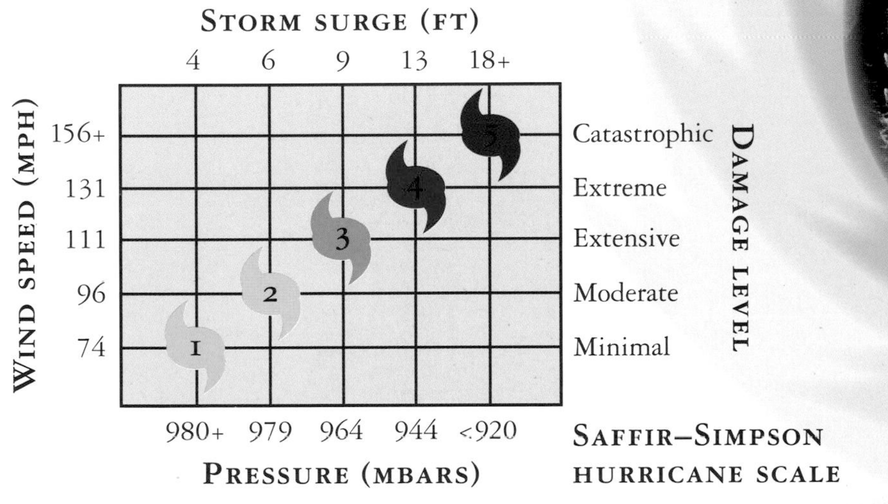

Tornadoes

Tornadoes are whipping whirlwinds of cloud that start on land. Although much smaller than hurricanes, they generate even more ferocious winds. They are ranked on the Fujita scale, with the worst (F-5) producing wind speeds of more than 320 km/h (200 mph).

Fujita scale

F-0 F-1 F-2 F-3 F-4 F-5

Very BIG

How big is Earth? The Milky Way? The Universe? They are so vast, our brains can't cope with the scale of them – the only way to grasp such huge sizes is to use maths.

Powerful numbers

Measuring big stuff uses big numbers, but writing these down can be a waste of time and paper. Instead, scientists use powers. A power shows how many times a quantity needs to be multiplied by itself. In the number 10^6 (said as "ten to the power of six"), 6 is the power. It's a quick way of writing $10 \times 10 \times 10 \times 10 \times 10 \times 10$ (1 million, or the number 1 followed by six zeros). Numbers that aren't simple multiples of 10 are written differently: 2 million is 2×10^6, and 7,654,321 is 7.654321×10^6.

$$10^3 = 1000$$

$$10^6 = 1,000,000$$

$$10^9 = 1,000,000,000$$

$$10^{12} = 1,000,000,000,000$$

THE COSMIC RULER

TALLEST BUILDING
The Burj Khalifa skyscraper in Dubai is 828 m tall (8.28×10^2 m).

MOON SIZE
The diameter of Earth's moon is 3477 km.

LONGEST RIVER
The Nile flows for 6695 km, from Rwanda to Egypt.

FURTHEST MOTORCYCLE RIDE
Emilio Scotto covered 735 million metres (and 214 countries) in his 10-year ride.

10^3 10^4 10^5 10^6 10^7 10^8 10^9 10^{10} 10^{11} 10^{12} 10^{13} 10^{14}

METRES

HIGHEST MOUNTAIN
Mount Everest in the Himalayas is 8848 m tall – around 10 times taller than the Burj Khalifa skyscraper.

JUPITER SIZE
Jupiter's diameter is 11 times that of Earth.

DISTANCE TO THE SUN
149,597,887,500 m (1 astronomical unit).

SOLAR SYSTEM SIZE
If Earth were as small as a pea, you could stroll across the 12-trillion-metre wide Solar System in an hour.

FURTHEST NONSTOP FLIGHT
On its first flight, a fledgling (young) swift may spend up to four years in the air, eating and sleeping on the wing. It flies around 800,000 km (8×10^8 m) without stopping.

SIZE OF EARTH
Earth's diameter (the width across the middle) is 12,756,000 m.

Huge units

When dealing with vast distances in space, metres simply aren't big enough, so scientists use a different set of measures: astronomical units (AU), light years, and parsecs. One AU is the distance from Earth to the Sun, and a light year is the distance light travels in one year. When a telescope shows us a view of something 10 light years away, that view happened ten years ago – it's taken ten years for the image to reach us.

Astronomical unit = 1.5×10^{11} m

Light year = 9.46×10^{15} m

Parsec = 3×10^{16} m

KILO PARSEC = 3×10^{19} m

MEGA PARSEC = 3×10^{22} m

Milky Way

Millions of stars, including the Solar System, make up the Milky Way. This galaxy stretches 100,000 light years from one side to the other.

Galaxy cluster

The Milky Way is just one of many galaxies in the Universe. Together with its neighbouring galaxies, they form a cluster (called the Local Group) that's 6 million light years wide.

Voids

There are huge holes in space where there are no stars, gas, or any other matter. These are called voids, and the biggest void discovered so far is almost 1 billion light years across. No one knows why the hole is there...

10^{15} 10^{16} 10^{17} 10^{18} 10^{19} 10^{20} 10^{21} 10^{22} 10^{23} 10^{24} 10^{25} 10^{26}

Orion nebula

This huge cloud made of dust and gas is 30 light years, or 280 quadrillion metres, wide.

How far can you see?

It's probably further than you think. Can you see stars on a clear night? They're light years away! The furthest we can see with the naked eye is usually the Andromeda Galaxy, 2.5 million light years away. Some people can even see the Triangulum Galaxy 3.14 million light years away.

The edge of the Universe?

How big is the biggest thing known to humankind: the Universe? Some say that it's as big as it is old, which would make it around 13.7 billion light years big. But that's not quite the whole story, because space itself is expanding. Taking that into account, the very furthest objects we can see are 46.5 billion light years away (4.4×10^{26} m). And that's just just the "observable" Universe – there could be a lot more beyond the reach of our telescopes. Nobody really knows how big the Universe is.

TO INFINITY and beyond...

Very SMALL

In the past, philosophers used to argue over how many angels could dance on the head of a pin, which was about the tiniest thing anyone could see. Now scientists routinely measure things 10 million times smaller.

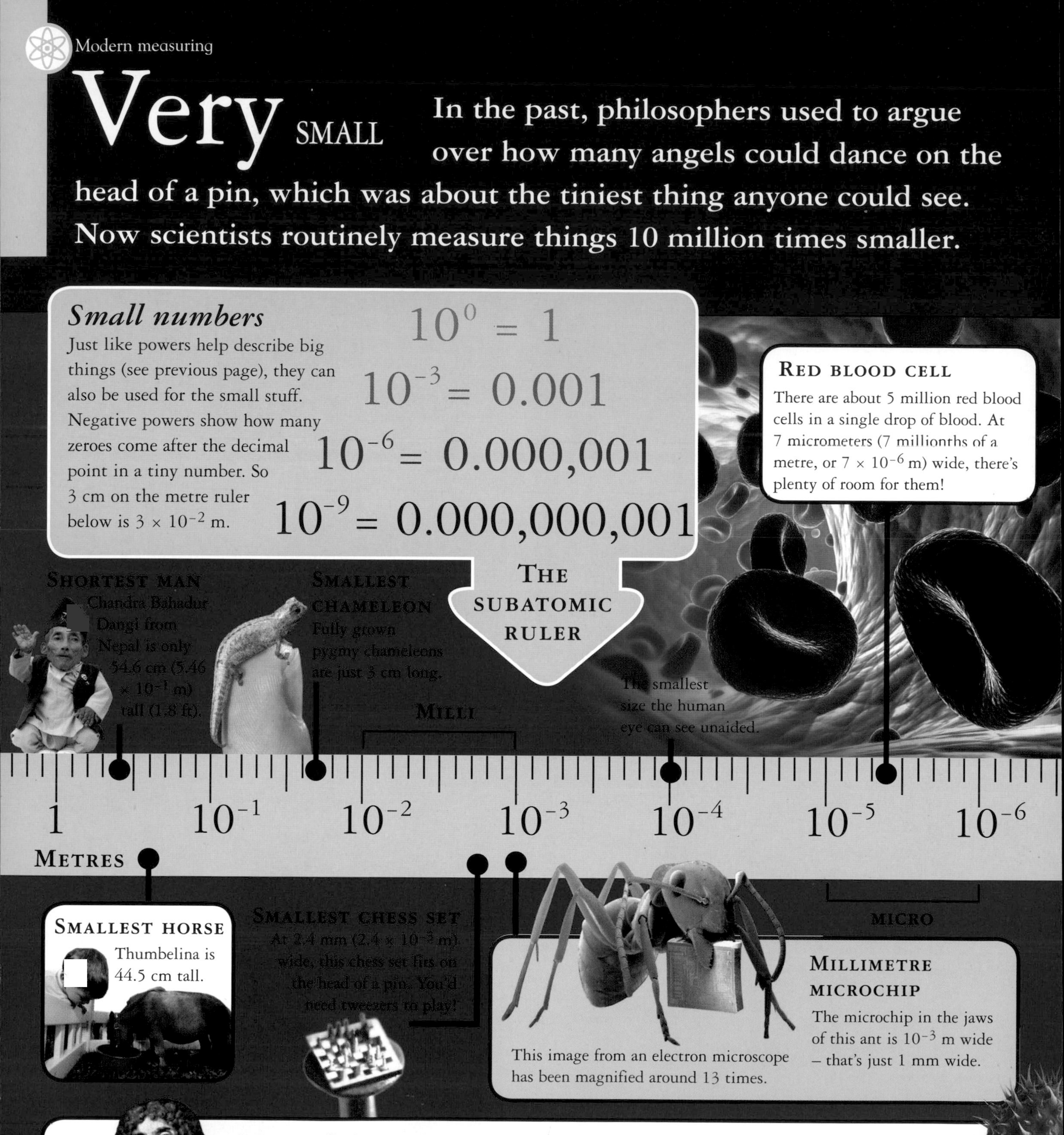

Small numbers

Just like powers help describe big things (see previous page), they can also be used for the small stuff. Negative powers show how many zeroes come after the decimal point in a tiny number. So 3 cm on the metre ruler below is 3×10^{-2} m.

$$10^{0} = 1$$

$$10^{-3} = 0.001$$

$$10^{-6} = 0.000,001$$

$$10^{-9} = 0.000,000,001$$

RED BLOOD CELL

There are about 5 million red blood cells in a single drop of blood. At 7 micrometers (7 millionths of a metre, or 7×10^{-6} m) wide, there's plenty of room for them!

SHORTEST MAN

Chandra Bahadur Dangi from Nepal is only 54.6 cm (5.46×10^{-1} m) tall (1.8 ft).

SMALLEST CHAMELEON

Fully grown pygmy chameleons are just 3 cm long.

SMALLEST HORSE

Thumbelina is 44.5 cm tall.

SMALLEST CHESS SET

At 2.4 mm (2.4×10^{-3} m) wide, this chess set fits on the head of a pin. You'd need tweezers to play!

MILLIMETRE MICROCHIP

The microchip in the jaws of this ant is 10^{-3} m wide – that's just 1 mm wide.

This image from an electron microscope has been magnified around 13 times.

Men of the microscope

Dutchman Anton van Leeuwenhoek (1632–1723) made the first scientific studies of the very small. He enjoyed scraping out and peering at gunge from old people's teeth and was one of the first people to see bacteria. He's often called the father of microscopy, but he actually used a glass bead magnifying glass. Englishman Robert Hooke (1635–1703) did use microscopes. He's best known for his book *Micrographia*, featuring sketches of animals under the microscope (including ants, which wouldn't keep still so he glued their feet down).

Hooke's sketch of a flea

Tiny units

Measuring tiny things in metres just isn't practical, so we use smaller measures instead. A pin-head is about two-thousandths of a metre wide, which is 2 mm or 2 million nanometres. To see nanoscopic things (such as atoms), scientists need electron microscopes. These use electron beams rather than light beams, so you can see things up to 1000 times smaller than can be seen through regular microscopes.

1 CENTIMETRE (cm)= 0.01 m = 10^{-2} m

1 MILLIMETRE (mm) = 0.001 m = 10^{-3} m

1 micrometre (µm) = 0.000,001 = 10^{-6} m

1 nanometre (nm) = 0.000,000,001 = 10^{-9} m

1 picometre (pm) = 0.000,000,000,001 m = 10^{-12} m

1 femtometre (fm) = 0.000,000,000,000,001 m = 10^{-15} m

1 yoctometre (ym) = 0.000,000,000,000,000,000,000,001 m = 10^{-24} m

1 Planck length = 0.000,000,000,000,000,000,000,000,000,000,000,016 m = 1.6×10^{-35} m

Nanotechnology

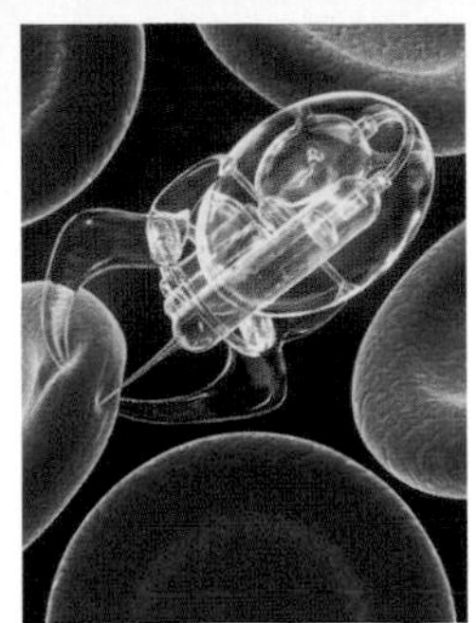

Once we can see and manipulate individual atoms, we may be able to start using them like building blocks in a nanoscopic construction set. In theory, with absolute control over atoms, we could build objects flawlessly, atom by atom. Perhaps in the future we could make nanobots – robots small enough to swim inside our blood vessels, repairing damage and killing diseases.

Neutron

Electron

Smallest radio

You'd have trouble plugging headphones into the world's smallest radio, which fits into a tube 0.00001 mm (10 nm) wide.

Fingernail growth

A fingernail grows 0.5 mm a week, which is about 1 nanometre (8×10^{-10} m) a second.

Helium atom

The radius (distance from centre to outer edge) of a helium atom is about 30 picometres. Trillions can fit on the head of a pin.

Proton

Around 1 millionth of a nanometre.

10^{-7} 10^{-8} 10^{-9} 10^{-10} 10^{-11} 10^{-15}

Nano

Pico

Atom art

In the 1990s, scientists working at the computer company IBM used a powerful electron microsope to nudge 35 xenon atoms into an IBM logo 5 nm tall.

Cold virus

The germs that give you a cold are 20 nanometres wide.

How small can you get?

There's a limit to how small something can get and still meaningfully exist. The tiniest measurement we have is called the Planck length (named after German physicist Max Planck). There's nothing that can possibly be smaller than a Planck length, which is around 10^{20} times smaller than a proton... well, apart from an electron. Or a quark. Or a lepton. All these "elementary particles" are thought to be the size of a "point", and as a point has no dimensions, these particles arguably have **ZERO SIZE!**

Planck length 1.6×10^{-35}

Electron 0?

Weird and wonderful

What you would use a *jerk* to measure? How about a *googol*, a *mickey*, a *garn*, or a *smoot*? Read on to find out about the world's weirdest and most wonderful units of measurement.

Googol

In 1938, mathematician Edward Kasner invented the googol. His 8-year-old nephew came up with the name. It's a really, really big number: 1 followed by 100 zeroes. The Google search engine was named after it.

Volume & purity

Carat (purity)

Why are some items made from gold more expensive than others? It's because the purity of gold, measured in carats, can vary a great deal. Pure gold is 24 carats, but 18-carat gold contains 18 parts gold and 6 parts other metals, making it only 75 per cent pure.

Sydharb

Australians use this unit of volume to measure water. One sydharb is the amount of water in Sydney Harbour, which is about 500 billion litres (900 billion pints).

Olympic-size swimming pool

An Olympic-size swimming pool is 50 m (164 ft) long × 25 m (82 ft) wide × 2 m (6.5 ft) deep. The huge unit of volume is handy for describing vast amounts. For instance, the UK produces enough rubbish to fill an Olympic-size swimming pool every four minutes.

Barn

This unit of area was born when a scientist joked that the nucleus of a uranium atom is "as big as a barn". In fact, one barn is very, very, very tiny indeed: 0.0000000000000000000000000001 square metres, to be precise.

Hmm... this isn't a very hygienic measurement.

Mouthful

The mouthful is about 28 ml and was once used to measure small volumes... yuk!

Speed & power

Horsepower

In the days of horse-drawn carts, people measured pulling power in number of horses. Oddly, we still rate cars and lorries in "horsepower" today. But not many people use the less well known unit "donkey-power". One donkeypower, in case you're interested, is a third of one horsepower.

Move it you stupid ass!

Knot

There's a good reason why knots – the units used to measure the speed of boats – are called knots. Sailors used to measure the speed of a ship by throwing a barrel tied to a knotted rope overboard. Using an hourglass, they counted how many knots floated past in a measured time period. One knot = 1.85 km/h – 1.15 mph (which is knot a lot).

The speed of light

The fastest thing in the Universe is the speed of light. It's impossible for anything to go faster – the laws of physics say so. Light travels through space at about 1 billion km/h (671,000,000 mph), which is fast enough to go right round Earth 7 times in a second. Pretty nippy!

Jerk

Woah!! Ever feel the jerk of a sports car when it suddenly accelerates? Engineers define "jerk" as the rate of change of acceleration and measure it in metres per second cubed.

measurements

googol
$= 10^{100}$

Scoville scale

The Scoville scale is what we use to measure the "heat" of chilli peppers. Watch out for the really hot ones!

- Bell pepper – 0
- Jalapeno – 2500
- Cayenne pepper – 30,000
- Habanero pepper – 200,000
- Carolina Reaper – 1,569,300 (the world's hottest chilli pepper)

Size

Cubit

This is the oldest known unit of length and was used in ancient Egypt. It's the length of a man's arm from elbow to the tip of his middle finger.

Barleycorn

This Anglo-Saxon unit was the length of a barley grain. In medieval Britain, three barleycorns made an inch (2.5 cm).

Digit and span

1. Digit
2. Span

Klick

ONLY 10 MORE KLICKS TO SAIGON!

Klick is army slang for kilometre. The term became popular in the 1960s among American soldiers in Vietnam. It probably came about because soldiers thought it sounded cool.

Elephant

In the 19th century there were no A4, A5 or A6 paper sizes. Instead, paper came in sizes such as "foolscap" (42 × 37 cm/16½ × 14½ in) to "elephant" (71 × 58 cm/28 × 23 in). And if you really wanted to impress, you could write your essay on the largest size of writing paper available at the time: the double elephant.

Furlong

This old English unit was the distance a plough was pulled across a standard field – about 201 m (660 ft). The furlong was abolished in 1985, but it's still sometimes used today in horse races. Maybe that's because racehorses don't stay still fur-long!

Weight

Grain

The grain is a unit of weight based on the seeds of wheat, barley, or other cereal crops. It's long been used to weigh small, precious items from coins and bullets to gunpowder.

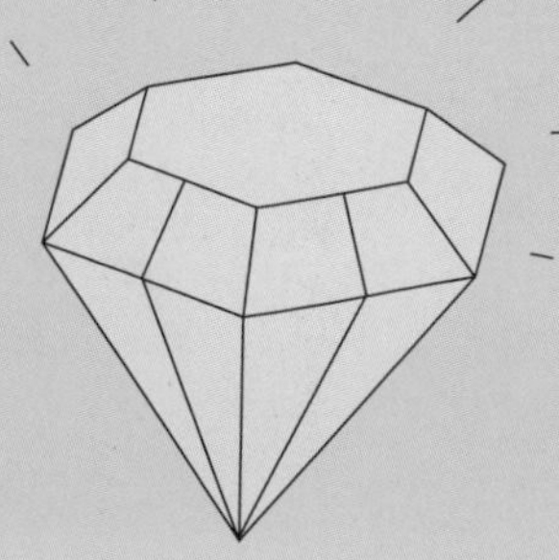

Carat (weight)

A measure of how heavy a diamond or other gemstone is. The word came from the Greek word for a carob seed, which was used as a standard weight in ancient Greece. It's now defined as 200 mg. Bling-bling!

Time

Atomus

In medieval times, the Latin word *atomus* meant "a twinkling of the eye" – the smallest amount of time imaginable. Nowadays it's defined as precisely ¹⁄₃₇₆ of a minute, or about 160 milliseconds. See you in an atomus!

Galactic year

This is the time it takes for the Solar System to make one complete orbit around the centre of our Milky Way galaxy. One GY = 250 million years. On the galactic timescale, the oceans appeared when Earth was 4 GY old and life began at 5 GY. Earth is currently 18 GY old – a mere teenager.

Jiffy

The length of this short unit of time depends on who you ask. Computer boffins define a jiffy as one tick of a computer's system clock (0.01 seconds). Physicists say a jiffy is the time it takes light to travel the width of one proton, making the jiffy an incredibly tiny 3×10^{-24} seconds.

Gigaannum (Ga)

The gigaannum (pronounced "gigga annum") is a billion years. Planet Earth formed 4.57 Ga (4.57 billion years) ago. Even more impressive – but much less useful – is the teraannum: 1 Ta is a trillion years, which is 70 times as long as the age of the Universe.

Beard-second

One beard-second is the length a man's beard grows in one second: 5 nanometres (0.000005 mm). This not-entirely serious unit is used only by atomic physicists to describe the tiny distances that atoms and subatomic particles move in. (Only they really know what they're talking about!)

Megaannum (Ma)

One megaannum (pronounced "mega annum") is a million years (1 Ma), which makes this unit handy for dealing with Earth's long history – what scientists call the "geological timescale". The dinosaurs bit the dust 65 Ma ago.

WAIT A MOMENT!

Moment

How long, exactly, are you asking someone to wait when you say "wait a moment"? The moment is a medieval unit of time equal to a fortieth of an hour, which is 1.5 minutes.

Computer

Mickey

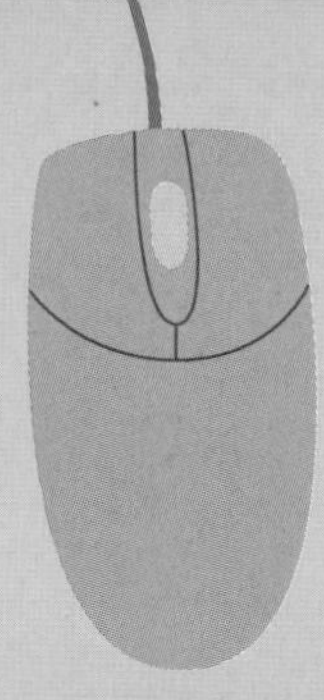

Named after the cartoon character Mickey Mouse, the mickey is the length of the smallest detectable movement of a computer mouse. It's about 0.1 mm (0.004 inches). Try saying this as fast you can: **"Mickey Mouse moved the mouse a mickey".**

A NYBBLE IS NOT ENOUGH. I WANT A BYTE!

Byte

We all know what megabytes and gigabytes are, but what exactly are bytes? Computers store all their information in binary code, which consists of a stream of 1s and 0s. Each 1 or 0 is called a "bit", and a collection of 8 bits is a "byte". The letter F, for example, is stored as one byte, made up of the bit pattern 01000110. A kilobyte is a thousand bytes, a megabyte is a million, a gigabyte is a billion, and a terabyte is a trillion.

Nybble

If you're peckish but don't want a whole meal, you might just have a nibble of something. In the world of computers, a "nybble" is exactly half a "byte". So what's a byte? Read on...

01000110

Named after people

I HAVE LAUNCHED A LOT OF SHIPS!

Warhol

Andy Warhol once said that "in the future everyone will be famous for fifteen minutes." So, a warhol is a measure of fame. 1 kilowarhol means being famous for 15,000 minutes or approximately 10 days.

Smoot

One smoot is defined as 1.7 m (5 ft 7 in), which was the height of US student Oliver Smoot in 1958. During a student prank at Harvard University, Smoot was used to measure Harvard Bridge. His pals laid him down on the bridge, drew a mark where his head was, and repeated the exercise all the way across. The length of the bridge was 364.4 smoots, plus or minus one ear. Smoot marks are still painted on the bridge to this day.

Millihelen

The millihelen is used to measure beauty. Helen of Troy – a stunningly beautiful queen of Greek mythology – had "the face that launched a thousand ships". The amount of beauty needed to launch one ship is a millihelen.

Garn

Sixty per cent of astronauts suffer from space-sickness while they are weightless in orbit. By far the worst case ever reported was that of Senator Jake Garn in 1985. He was so sick that his name is now used by NASA as a unit of measurement for space-sickness. One garn is the most sick you can get!

Miscellaneous

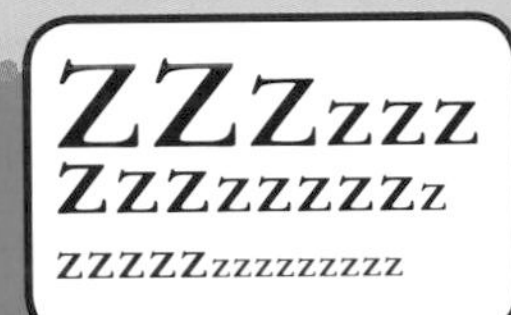

Apgar score

When you were born you were given an Apgar score. It's the first test you took! The Apgar score evaluates the health of newborns immediately after birth, based on their Appearance, Pulse, Grimace, Activity, and Respiration. It ranges from 0 to 10.

Hobo power

This is a measure of how bad something smells. It ranges from 0 (no smell) to 100 (lethal). A robust fart is about 13 hobo. At 50 hobo, the person doing the smelling would definitely vomit. Yuck!

Big Mac index

The Big Mac Index was invented by economists to compare the spending power of different currencies. For instance, if a Big Mac is £1 in the UK and $2 in the US, but the exchange rate is £1 = $1.50, then the British pound has more spending power and might be overvalued (which means it could tumble in value in the future).

Calorie

The calorie (also called kilocalorie) is used to measure how much heat energy food releases when it burns. The more energy the food contains, the more fattening it is. One kilocalorie is the energy needed make 1 kg of water 1°C warmer.

Flock

Ever wonder how many birds are in a flock of seagulls? A flock means 2 score or 40.

Decibel

We measure sound intensity in decibels, a scale named after telephone inventor Alexander Graham Bell. A 10-decibel increase is actually a tenfold increase in power, so a 40 dB sound is 1000 times more powerful than a 10 dB sound (but sounds only 8 times louder).

Baker's dozen

A baker's dozen is 13. This ancient measure dates back to 13th century England when bakers who were caught cheating customers were punished by having a hand chopped off with an axe. To guard against this unpleasant fate, bakers threw in an extra loaf for free when a customer bought a dozen (12). Best to be safe!

Free

The *METRIC* system

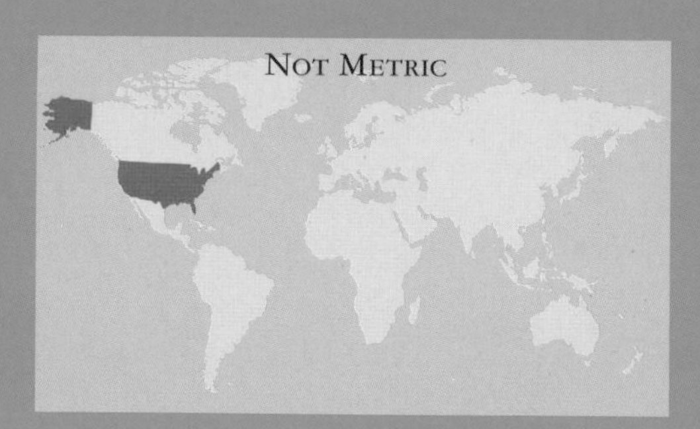

Nearly every country in the world uses the metric system for official measures. Having one system helps international trade: people making 10 mm screws in Peru can sell them to people wanting 10 mm screws in Switzerland – and the Swiss know the screws will be the exact size they need, as everyone uses the same standard measurements.

NOT METRIC

France invented the metric system more than 200 years ago, and it has since been adopted almost worldwide. The US is the only country officially not to be metric.

BEFORE METRIC...

...there were all sorts of different and complicated systems for measuring things. Look at length: there are 12 inches in a foot, 3 feet in a yard, 1760 yards in a mile, plus chains, furlongs, rods, spans, barleycorns, and ells. It all added up to a lot of seemingly random measures and awkward numbers that were not easy to work with.

Pah! Just one ell, mein Herr? That's so small!

I tell you sire, the fish was at least an ell long!

What the ell?

Even more confusing than the awkward numbers was the inconsistency of measures. Take the ell. When first used in England in the Middle Ages, it was based on the length of a man's arm, around 57 cm by today's terms. But it was later changed by Parliament to twice that length. At the same time, in Germany it was around 40 cm, but in Scotland, 95 cm. And in Switzerland alone there were 68 different actual lengths all referred to as an ell. Surely there had to be a better way...?

The *better* way

First developed in the 1790s, the metric system made measuring simple. Now called the International System of Units (or SI), it provides one set of consistent, easy-to-use units. The modern system has seven main units ("base units") from which we derive all other units (such as square metre for measuring area).

Ammeters measure electric current in amperes.

UNIT	SYMBOL	QUANTITY (WHAT IT'S USED TO MEASURE)
metre	m	length
kilogram	kg	mass
second	s	time
ampere	A	electric current
kelvin	K	thermodynamic temperature
mole	mol	amount of substance
candela	cd	luminosity (how bright something is)

THE SEVEN BASE UNITS OF THE METRIC SYSTEM

If you're wondering what's happened to the metric measures you learned in school, such as litres, tonnes, and degrees Celsius, don't worry. While they're not official SI units, they're accepted by the system.

Decimal DELIGHT

The big advantage of the metric system is that it's a decimal system: the units can easily be made bigger or smaller by multiplying by a factor of 10. For example, you can measure an ant not in metres, but thousandths of a metre – far more appropriate. Even more handily, multiples are identified by prefixes. So instead of saying the ant is 9 one-thousandths of a metre long, it's simply 9 millimetres long.

PREFIX	MEANING	SYMBOL	WRITTEN AS
tera	Greek for monster	T	1,000,000,000,000
giga	Greek for giant	G	1,000,000,000
mega	Greek for big	M	1,000,000
kilo	Greek for thousand	k	1000
hecto	Greek for hundred	h	100
deka	Greek for ten	da	10
deci	Latin for tenth	d	0.1
centi	Latin for hundredth	c	0.01
milli	Latin for thousandth	m	0.001
micro	Greek for small	μ	0.000,001
nano	Greek for dwarf	n	0.000,000,001
pico	Spanish for tiny bit	p	0.000,000,000,001

Deadly *ERRORS*

The USA is the only country not to officially adopt the metric system (though it is widely used in science and industry). Instead, they have "customary units". Using two systems is not just confusing, it can be dangerous. In 1983, a Boeing 767 was refuelled with 22,600 **lb** of fuel. But it should have been 22,600 **kg** – more than twice as much. Unsurprisingly the jet ran out of fuel; it was only the pilot's skill in gliding the plane into land that saved the lives of those on board. Even scientists are not immune: NASA crashed a Mars orbiter because one team measured in metric and the other in US customary units.

The size of a centimetre has not changed *in more than 200 years.*

Setting standards

In 1792, in the middle of the French Revolution, two French astronomers measured the distance between Dunkirk and Barcelona and then worked out the distance from the North Pole to the equator. They called it 10 million metres. Dividing this distance by 10 million set the length of a metre, which became the first unit in the metric system. But how would the average person know how long a metre should be? They needed a guide. So, in 1799, two platinum standards were made – models that showed the official length of a metre and mass of a kilogram.

The kilogram standard was replaced in the 1880s, and the new one is kept under glass in a vault in Paris. To check a kilogram weight has an exact mass of 1 kg, you have to compare it to the standard. But the mass of the standard kilogram has shrunk by about 30 μg (30 millionths of a gram) since it was made!

Answers

MEASURING LAND (page 21) PUZZLE

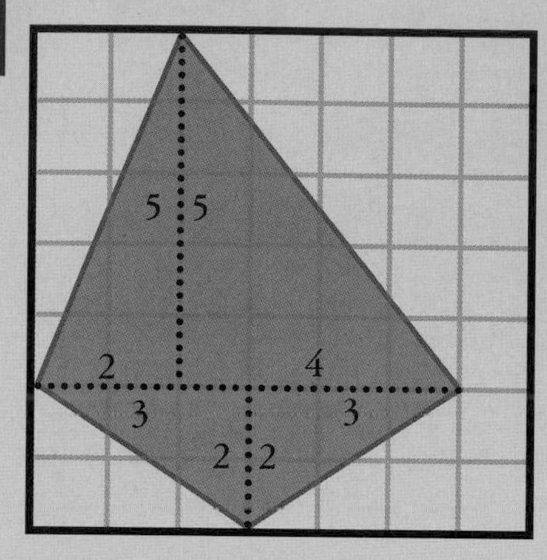

Divide the shape into right-angled triangles. Work out the area of each triangle by working out the area of each rectangle (multiply the length by the width) and halving it. Then add them together.

$\frac{5\times2}{2}=5 \quad \frac{5\times4}{2}=10 \quad \frac{3\times2}{2}=3 \quad \frac{3\times2}{2}=3$

$5 + 10 + 3 + 3 = \mathbf{21\ cm^2}$

WHY MEASURE ANY BODY? (page 36–37)

The statement is true. Most people have two legs, but the average number of legs is less than this. Among the billions of people on Earth, there are many thousands who have only one leg or no legs. Suppose Earth's population is 6700 million, and there are a million people with only one leg and a million with none.

The total number of legs is:

$(6{,}698{,}000{,}000 \times 2) + 1{,}000{,}000 = 13{,}397{,}000{,}000$

The total number of people is:

6,700,000,000

The average (mean) number of legs is:

$$\frac{13{,}397{,}000{,}000}{6{,}700{,}000{,}000} = 1.9995$$

So if you have two legs, you have more than average!

WEIGHING UP (page 43) FRUIT PUZZLE

If 1 orange + 1 plum = 1 melon
And 1 orange = 1 plum + 1 banana
And 2 melons = 3 bananas
How many plums equal 1 orange?

Answer

From statements 1 and 2,
1 melon = 2 plums + 1 banana
So 2 melons = 4 plums + 2 bananas
Also, 2 melons = 3 bananas
So 4 plums = 1 banana
So 5 plums = 1 orange.

HEAVY HEAD PUZZLE

1. Stand a bucket in a large tray and fill the bucket with water right up to the brim.
2. Lower in your head so that it's completely immersed, displacing its own volume of water.
3. The displaced water has the same volume as your head. Your head and the displaced water also weigh about the same because their densities are about equal. So if you weigh the displaced water you'll get a pretty good answer for the weight of your head!

Acknowledgements

Dorling Kindersley would like to thank Ria Jones for help with picture research.

The publisher would like to thank the following for their kind permission to reproduce their photographs:

(Key: a-above; b-below/bottom; c-centre; f-far; l-left; r-right; t-top)

9 Corbis: Mike Agliolo (cl). Science Photo Library: National Institute of Standards and Technology (NIST) (bl). 10 Corbis: Richard Bryant / Arcaid (c); Jose Fuste Raga (cr). Getty Images: Ron Dahlquist (ca); Don Klumpp (fcr). 11 Corbis: Bettmann (br). Getty Images: Jonny Basker (bl). 12 Corbis: Werner Forman (cr). Getty Images: Garry Gay (cl); Image Source (bc). 13 Mary Evans Picture Library: (tl). Science Photo Library: Sheila Terry (cl). 14 Science Photo Library: Gary Hincks (br/Sun). 15 Science Photo Library: Gary Hincks (br). 16 Science Photo Library: Mark Garlick (br). 17 Science Photo Library: Mark Garlick (br). 19 NASA: Satellite Imaging Corporation (bl). 21 Corbis: Werner Forman (tl). 22 Science Photo Library: (bc). 24 Science Photo Library: Sheila Terry (bl). 25 Getty Images: World Perspectives (tl). 31 Corbis: Yann Arthus-Bertrand (cr). 32 Mary Evans Picture Library: (bl). 34 Getty Images: Image Source (bc). 35 Corbis: Hanan Isachar (bl). Getty Images: Garry Gay (tl). Science & Society Picture Library: Science Museum (br). 36 Corbis: David Cumming (ca). TopFoto.co.uk: The British Library / HIP (cl). 37 Getty Images: Garry Gay (tl). 38 DK Images: Science Museum, London (bl). 38-39 Corbis: Roger Ressmeyer (tc). Getty Images: Doug Armand (c). 39 Corbis: Bettmann (cl); Jack Hollingsworth (tc). DK Images: Science Museum, London (cr). 40 Corbis: Art on File (c). iStockphoto.com: Joachim Angeltun (crb); edge69 (cr). 41 Corbis: Roger Ressmeyer (cra). 42 Corbis: Hoberman Collection (crb) (br). 43 DK Images: Natural History Museum, London (tc). Science Photo Library: (bl). 44 Corbis: (clb); Mike Agliolo (cl). 45 Corbis: Bettmann (tl). 46 Science Photo Library: Sheila Terry (br). 47 Science Photo Library: Maria Platt-Evans (tl) (bl). 50 DK Images: National Maritime Museum (br) (cra/Kepler). Science Photo Library: Maria Platt-Evans (cra/Galileo) (cra/Newton); Sheila Terry (bl). 51 Corbis: Tim Kiusalaas (cl). 52 Alamy Images: North Wind Picture Archives (cla). Corbis: Paul Almasy (br). 53 Corbis: Mike Agliolo (br); Michael Nicholson (tr). 54 Corbis: (bl). 54-55 Corbis: (c/Background). 55 Corbis: Bettmann (cb). DK Images: National Maritime Museum (ca) (bc) (c). National Maritime Museum, Greenwich, London: (tc). 56 Science Photo Library: (cl) (br); Royal Astronomical Society (fbr). 57 Corbis: Hulton-Deutsch Collection (br); Roger Ressmeyer (t). DK Images: NASA (tc). 58 Alamy Images: Classic Image (cb/Columbus). The Bridgeman Art Library: Royal Geographical Society, London, UK (tl) (cb/Boat). Corbis: Bettmann (cla) (bl). iStockphoto.com: Julien Grondin (Background). 59 Alamy Images: Classic Image (tl). 60 Getty Images: Ted Kinsman (tl). Science Photo Library: (tr). 61 Corbis: Randy Faris (bl); Martin Gallagher (tl). 62 Getty Images: Ted Kinsman (br). iStockphoto.com: Ted Grajeda (bl). 63 Corbis: Hulton-Deutsch Collection (cr). iStockphoto.com: Ted Grajeda (tr). NASA: NASA, ESA and The Hubble Heritage Team (STScI/AURA) / J. Biretta (br). Science Photo Library: National Institute of Standards and Technology (NIST) (bc) (cr/Portrait). 64 Corbis: Bettmann (cl). Getty Images: Dougal Waters (cr/Hands). 65 Alamy Images: Elmtree Images (bc/Train). DK Images: NASA / Finley Holiday Films (br/Space Shuttle); Toro Wheelhorse UK Ltd (bl/Lawnmower). Getty Images: AFP (fbr/Earthquake); Andy Ryan (fbl/Runner). 67 Alamy Images: Realimage (crb). Corbis: Chris Collins (hair dryer); Martin Gallagher (br); Image Source (kettle); LWA-Stephen Welstead (fluorescent light bulb); Lawrence Manning (fridge) (microwave); Radius Images (laptop); Jim Reed (tr); Tetra Images (incandescent light bulb) (c). 68 Corbis: (crb); Lawrence Manning (crb/thermometer). NASA: (bc). 68-69 Science Photo Library: Sheila Terry (cb). 69 NASA: (bc). Wikimedia Commons: (crb). 70 Science Photo Library: Chris Butler (b). 70-71 Science Photo Library: Pekka Parviainen (t). 71 Science Photo Library: Eckhard Slawik (b). 72 Science Photo Library: (br). 74 Corbis: Bettmann (b). 75 DK Images: The British Museum (cr). Science Photo Library: Hank Morgan (cra). 76 Corbis: David Arky (bl/violin). Getty Images: (bl); Arctic-Images (tl). 77 Corbis: Randy Faris (cla). 78 Alamy Images: nagelestock.com (bl). Science Photo Library: Gregory Dimijian (cl). Chris Woodford: (clb). 79 Getty Images: (br). Science Photo Library: Lande Collection / American Institute Of Physics (tl); NOAO / AURA / NSF (bl). 80 Getty Images: Mads Nissen (bl). Science Photo Library: David A. Hardy (cra). 81 Corbis: Frans Lanting (tl). Getty Images: Paul & Lindamarie Ambrose (cr); Dr. Robert Muntefering (c). 82 Alamy Images: Arco Images GmbH (bc). Corbis: NASA/JPL-Caltech (crb); William Radcliffe/Science Faction (bl) (clb) (fcrb). Getty Images: AFP (fcla); Paul Joynson Hicks (cla); Travel Ink (fclb). 83 Corbis: Tony Hallas/Science Faction (bl); Myron Jay Dorf (tl). Getty Images: Jack Zehrt (bc). Science Photo Library: David Parker (br). 84 Corbis: Sarah Rice/Star Ledger (clb), See Li/Demotix (cl); Popperfoto (bl). Ben Morgan: (clb/chess set). Science Photo Library: Alexis Rosenfeld (cl); Andrew Syred (crb). Wikimedia Commons: (br). 84-85 Getty Images: 3D4Medical.com (ca). 85 Corbis: Matthias Kulka/zefa (fbl/virus). Getty Images: Image Source (cl); Gabrielle Revere (bl). Image originally created by IBM Corporation: (clb). Science Photo Library: Coneyl Jay (cla). 86 Getty Images: artpartner-images (cla); Erik Dreyer (br); Chad Ehlers (cl); FPG (cr). 88 Alamy Images: imagebroker; The Print Collector (cl). Corbis: Tetra Images (br). DK Images: Anglo-Australian Observatory (tc). Getty Images: De Agostini (cr). 89 Alamy Images: Classic Image (tl); The Print Collector (crb). Science Photo Library: Sheila Terry (bl). 90 Getty Images: David Muir (br). Science Photo Library: Andrew Lambert Photography (bl). 90-91 iStockphoto.com: Björn Magnusson (t). 91 Alamy Images: Mint Photography (cr). Getty Images: AFP (br). 93 Science Photo Library: (br)

BOO!

INDEX